Stormwater Design Toolkit

SUSTAINABLE STORMWATER UPDATE TO THE COMMUNITY REDEVELOPMENT AREA STORMWATER MASTER PLAN

January 2012

Dynamic   Innovative   Sustainable

GOLDEN GATE • HOBE SOUND • INDIANTOWN • JENSEN BEACH • PALM CITY • PORT SALERNO • RIO

Table of Contents

INTRODUCTION

Introduction..............................4
Purpose..................................4
Stormwater..............................5
Sustainability...........................6
Watershed...............................6
Establishment of the CRA............7
Development in Action..............7
Livability Principles...................8

TOOL BOX

How to Use.............................10
Categories..............................11
Pavement...............................12
Channeling.............................26
Storage..................................33
Filtration................................46

MODULES

How to Use.............................68
50' Roadway Section................69
80' Roadway Section................71
100' Roadway Section..............73

IMPLEMENTATION

Mixing Materials......................76
Next Steps..............................77
Acknowledgements..................78

Martin County Board of County Commissioners
Your County, Your Community

Introduction

I Introduction

CRA Goals

Martin County Community Redevelopment Agency (CRA) provides this report to:
- Inspire future development to include New Urban and Traditional Neighborhood Designs Principles
- Encourage future developments to utilize innovative, sustainable stormwater management strategies
- Demonstrate that over time these strategies are more cost effective than conventional suburban development
- Incorporate these strategies into Community Oriented Development Express code (CODEx) which is anticipated to be codified into Martin County Land Development Regulations by Summer 2012.

INTRODUCTION

Funding and support for this project is provided by South Florida Water Management District and a successful partnership with Martin County Redevelopment Agency. The aim of the project is to update the existing Martin County Redevelopment Area Stormwater Master Plan describing and demonstrating innovative engineering and best management practices. This update will provide additional sustainable elements that can be "modularized" to provide alternative approaches, urban strategies and innovative methodologies toward treating and managing stormwater. Various tools found within the project can be combined to create effective "treatment trains" designed to maintain and treat stormwater runoff. Each tool has the ability to be used as a piece of the development puzzle assisting a property owner to evaluate the capacity of a piece of land and the value of each tool's potential.

PURPOSE

This report is intended to be dynamic and applicable to a wide audience. It was intentionally written in plain language so that people of all ages and backgrounds could be inspired to learn more about stormwater management best practices and innovative strategies.

Throughout Florida you can see the effects of poor environmental stewardship. The Martin County CRA vision is to be a leader in Florida to encourage a built environment that reflects sustainability, ultimately resulting in a positive impact on the quality of life for its residents and continued opportunities for its businesses.

Martin County CRA seeks to finds a way to recognize and celebrate both the natural and the built environments, without one being at the expense of the other. The intent is to encourage the redevelopment areas to become more balanced where both environmental preservation and quality of life are of equal importance.

STORMWATER

Stormwater is precipitation that falls from the sky due to natural weather conditions onto the earth's surface. The hydrologic cycle of precipitation includes rainfall, evaporation/transpiration, surface runoff, interflow and base flow conditions. Pre-development or natural conditions allow stormwater to be intercepted by tree canopy, natural landscape, and begin to infiltrate the natural soil conditions and percolate into the groundwater. The volume of water that is not immediately utilized by the natural surroundings is classified as runoff. The runoff in pre-development condition is typically minimal and classified as balanced. This is due to the direct correlation between the natural growth of plants/trees and the local rainfall; i.e. high rainfall will see dense lush growth, low rainfall areas will see drought tolerant scrub. The natural cycle maintains a healthy groundwater table at a level to support naturally occuring water bodies.

The hydraulic cycle of precipitation of post-development changes drastically based on the amount of impervious or solid surface the rainfall encounters. A higher percent of impervious area results in a greater volume of runoff which negatively impacts the surrounding landscape and downstream water. Negative impacts include drought conditions from evaporation, flash flooding and pollution, all of which are due to the lack of infiltration and percolation. Groundwater tables diminish and the level of natural water bodies decrease or disappear. To combat the negative impacts, development tools are implemented to mimic natural conditions. These tools increase infiltration and percolation, the storage of water, and the evaporation/transpiration at a rate similar to natural conditions. The term "treatment train" is used to describe this implementation of tools in a step by step manner, recreating natural conditions.

By taking innovative steps to increase the implementation of tools to recreate natural stormwater treatment, we can increase development in a smart way and decrease the pollution and negative effects to the watershed.

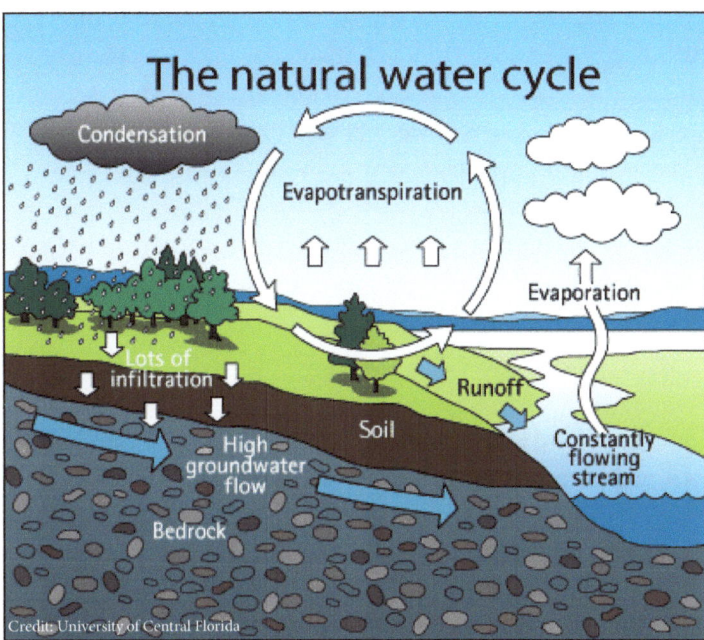

Natural Water Cycle: *This illustration shows the natural progression and entire life cycle for stormwater in a natural habitat untouched by man.*

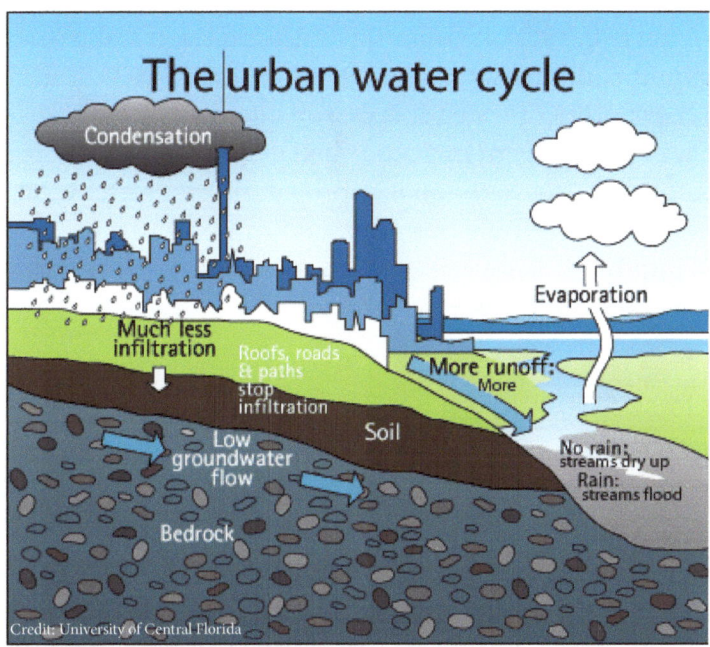

Urban Water Cycle: *This illustration shows how the same life cycle for stormwater when urban elements are introduced in the cycle. This cycle is the focus of this report.*

STORMWATER DESIGN TOOLKIT
Martin County Community Redevelopment Agency

SUSTAINABLITY

Sustainability is an underlying theme. The CRA encourages development using New Urbanist planning principles, with a focus on creating a place that fits into the landscape, fulfills the needs of a community and improves the future of the natural world. As a result, the CRA will be a prime example of how development can be accomplished in a manner and scale that respects and protects the intrinsic natural qualities of the site and the region and celebrate life in a way not usually available in conventional suburban developments.

WATERSHED

Martin County CRA is located in Watershed #6 and #12 St Lucie Loxahatchee. A watershed is simply the geographic area through which water flows across the land and drains into a common body of water, whether a stream, river, lake, or ocean. Much of the water comes from rainfall and the stormwater runoff. The quality and quantity of stormwater is affected by all the alterations to the land-agriculture, roadways, urban development, and the activities of people within a watershed. Watersheds are usually separated from other watersheds by naturally elevated areas.

Because the surface water features and stormwater runoff within a watershed ultimately drain to other bodies of water, it is essential to consider these downstream impacts when developing and implementing water quality protection and restoration actions. Everything upstream ends up downstream. It is important to remember that everyday activities can affect downstream waters.

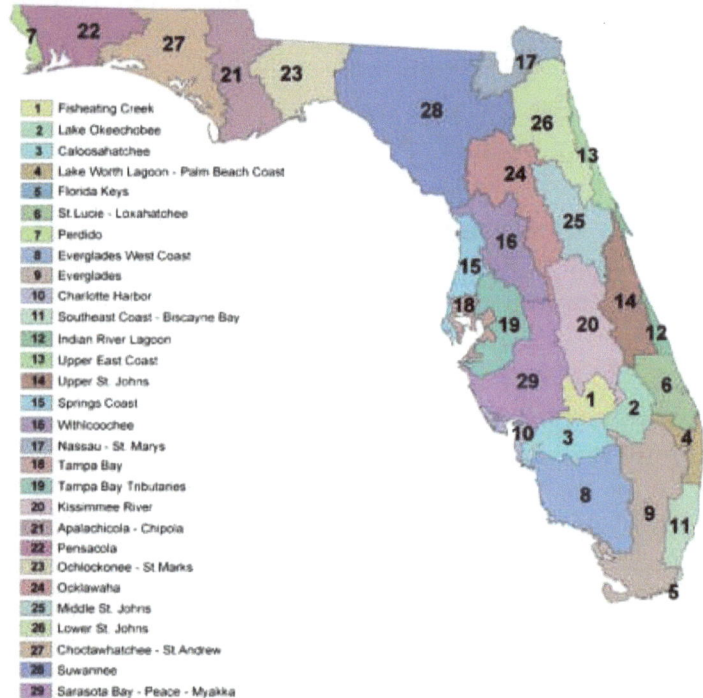

Watershed: *Martin County is located in Watershed #6 St Lucie Loxahatchee*

DEVELOPMENT IN ACTION

The CRA offers opportunities for both infill and new development. These pose very different challenges. Existing communities which were built before stormwater management was mandated must be retrofitted to evolve into areas with adequate treatment and storage. Newer communities within the CRA must manage their stormwater but be mindful of how each strategy could affect surrounding, existing development within the CRA.

Implementation: *Recent stormwater and streetscape improvements.*

Martin County Board of County Commissioners
Your County, Your Community

SUSTAINABLE STORMWATER UPDATE

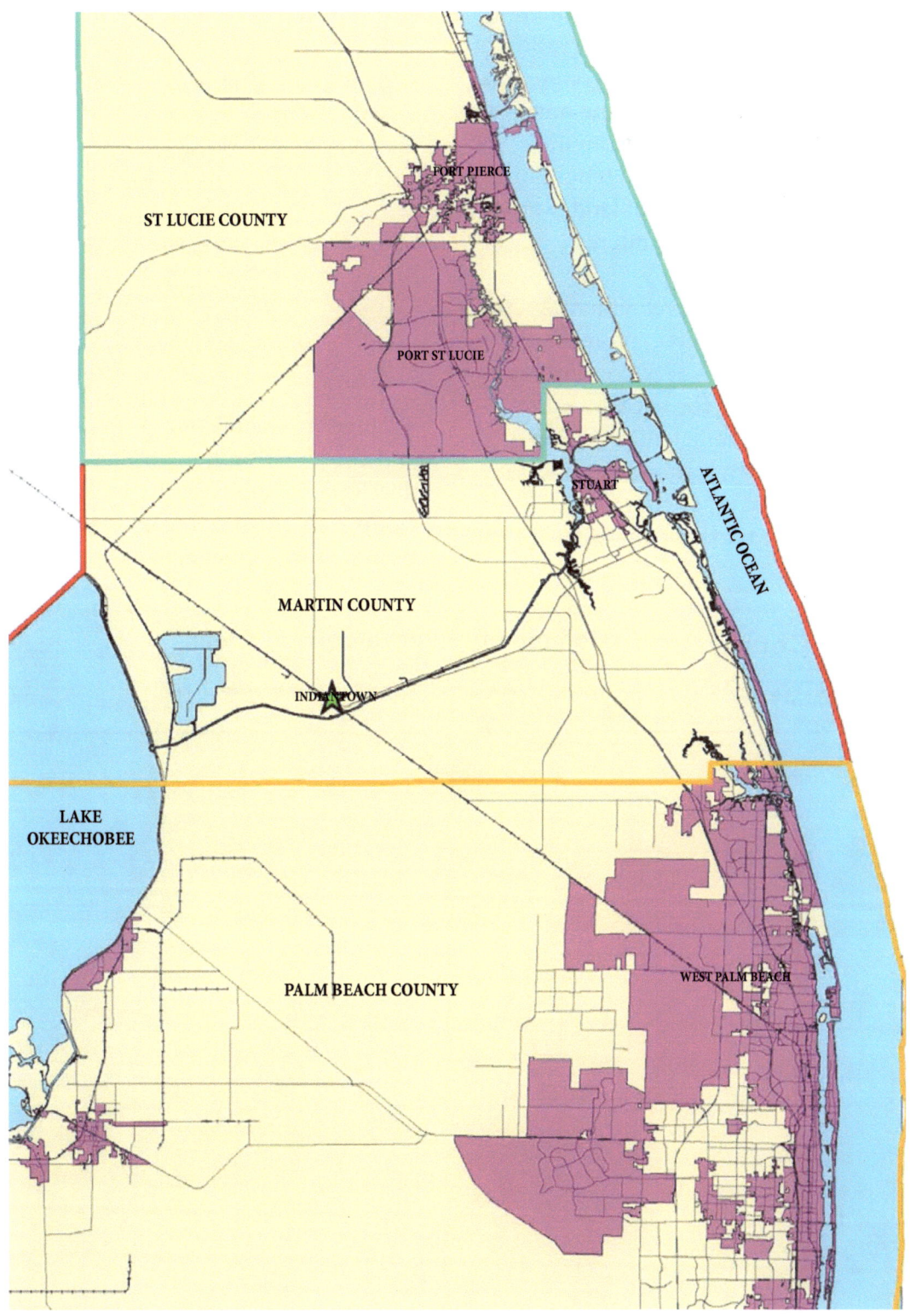

Regional Locator: *Martin County is located in Watershed #6 St Lucie Loxahatchee, and is geographically between Lake Okeechobee and the Atlantic Ocean on the Treasure Coast.*

Martin County Community Redevelopment Agency
Redevelopment in Action

STORMWATER DESIGN TOOLKIT
Martin County Community Redevelopment Agency

ESTABLISHMENT OF THE CRA

At the beginning of this century, the Martin County Board of County Commissioners recognized the unfulfilled potential of seven targeted areas of Martin County. They designated them as redevelopment areas which allows revitalization and preservation of once viable neighborhoods and thriving commercial activity centers. By taking advantage of the numerous tools afforded by State Statute, these community redevelopment areas are encouraged to redevelop by employing the most appropriate use of land, water, and resources consistent with the public interest.

Through the Martin County Redevelopment Agency (CRA), Martin County can preserve, promote, protect, and improve the public health, safety, comfort and general welfare of these areas. The CRA intends to facilitate the adequate and efficient provision of transportation, potable water, wastewater collection and treatment, housing and other community services in areas that have traditionally suffered from disinvestment.

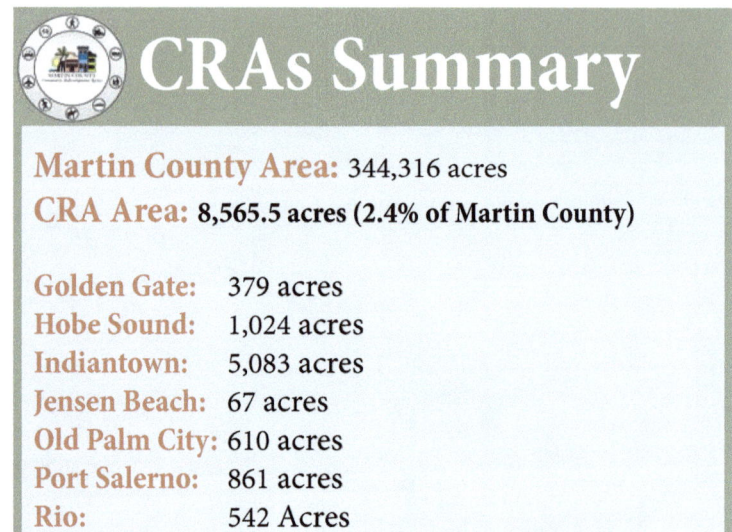

CRAs Summary

Martin County Area: 344,316 acres
CRA Area: 8,565.5 acres (2.4% of Martin County)

Golden Gate:	379 acres
Hobe Sound:	1,024 acres
Indiantown:	5,083 acres
Jensen Beach:	67 acres
Old Palm City:	610 acres
Port Salerno:	861 acres
Rio:	542 Acres

This report will encourage and assure cooperation and coordination in planning, redevelopment, and development activities between and among Martin County, its citizens, regional agencies, and state and federal government agencies.

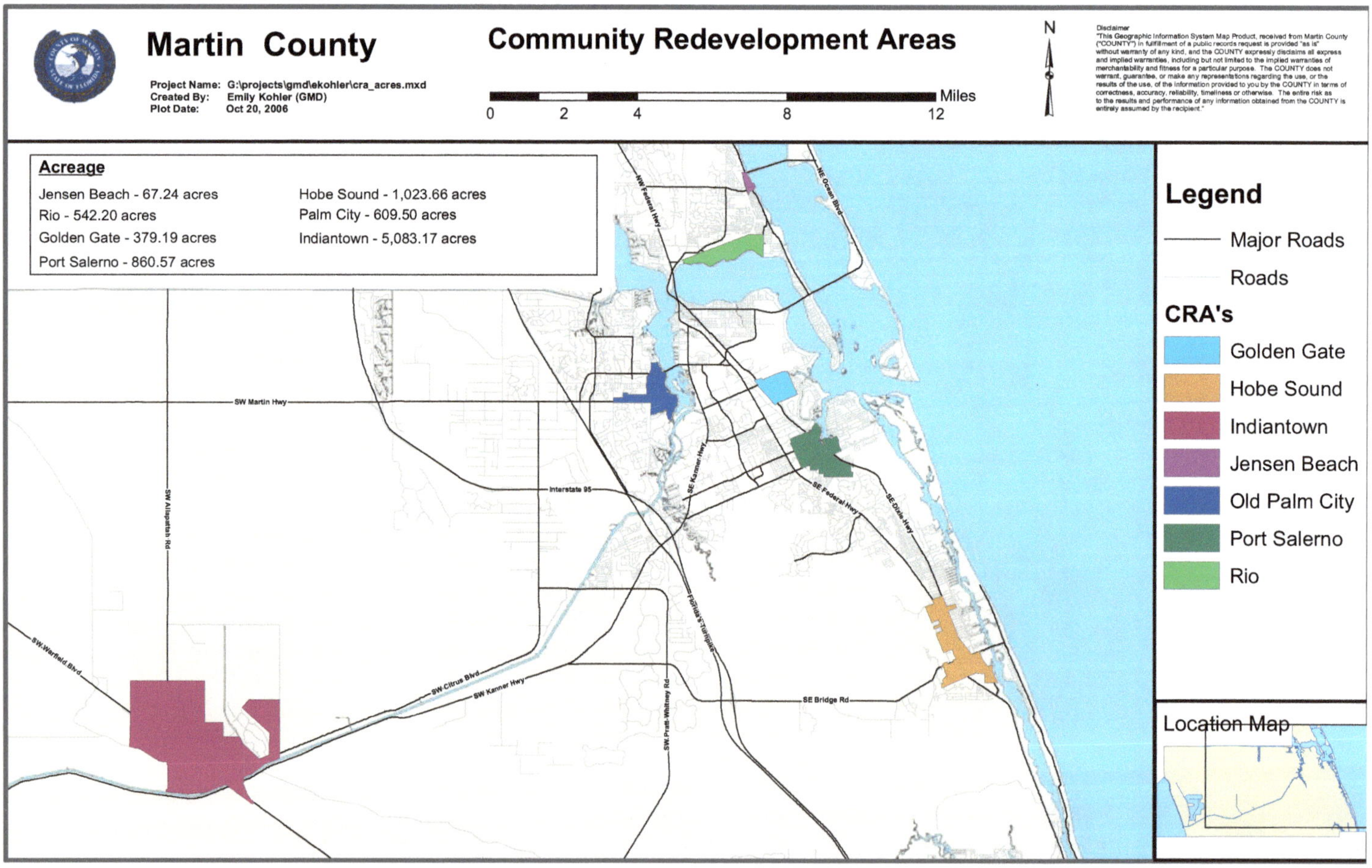

Martin County Board of County Commissioners
Your County, Your Community

LIVABILITY PRINCIPLES

The Martin County Community Redevelopment Agency supports interagency partnerships that support principles of livability as outline from *The Partnership for Sustainable Communities:*

1. PROVIDE MORE TRANSPORTATION CHOICES
By utilizing these strategies, the CRA developments will provide safe alternatives to conventional automobile travel.

2. PROMOTE EQUITABLE, AFFORDABLE HOUSING
Better utilizing existing land offers a more efficient way to develop that encourages a variety of housing types.

3. ENHANCE ECONOMIC COMPETITIVENESS
By developing in an efficient way, businesses can prosper and develop in an affordable way.

4. SUPPORT EXISTING COMMUNITIES
The CRA targets local funding toward existing communities – through strategies such as dynamic stormwater management to complement existing infrastructure, improve the efficiency of public works investments, and safeguard rural landscapes.

5. COORDINATE POLICIES AND LEVERAGE INVESTMENT
Provide stormwater management best practices that complement the existing environment and encourage innovative stormwater management strategies that maximize the existing developable land.

6. VALUE COMMUNITIES AND NEIGHBORHOODS
The open spaces, natural landscapes and environmental stewardship all contribute to the strong sense of place. Efficient stormwater management allows for more open space so the community can develop attractive places for people to gather, meet each other, and celebrate the great collective moments of life.

How to Use

STORMWATER DESIGN TOOLKIT
Martin County Community Redevelopment Agency

How to Use

TOOL BOX

Each strategy is summarized on individual "tear sheets" for easy access and reference. Each sheet includes a description of the strategy, its unique qualities such as stormwater management and/or treatment; the cost to construct and the cost to maintain. In some cases an investment up front can be offset significantly by lower on-going maintenance costs or the ability to increase the density on the property.

Each sheet also includes a table summarizing the Construction and Maintenance costs defined as low, moderate or high referencing a "standard cost" of conventional development; the stormwater values of water quality and storm control referencing the qualitative and quantitative efficiency of the tool; and the appropriateness of the tool based upon water table defined as the naturally occurring level of water beneath the surface of the soil or existing ground cover; and soil permeability defined as the rate water can flow through which is described as high for clean sands and low for soils with high organics or clays.

Rock Aggregate			
Cost Factor			
	L	M	H
Construction Cost	$		
Maintenance Cost	$		
Stormwater Value			
	L	M	H
Water Quality		X	
Storm Control	X		
Water Table and Soil Types			
	L	M	H
Water Table	X	X	X
Soil Permeability		X	X
L - Low			
M - Medium			
H - High			

- A. Title
- B. Table
- C. Images, Diagrams, and Applications
- D. Description
- E. Qualities
- F. Costs

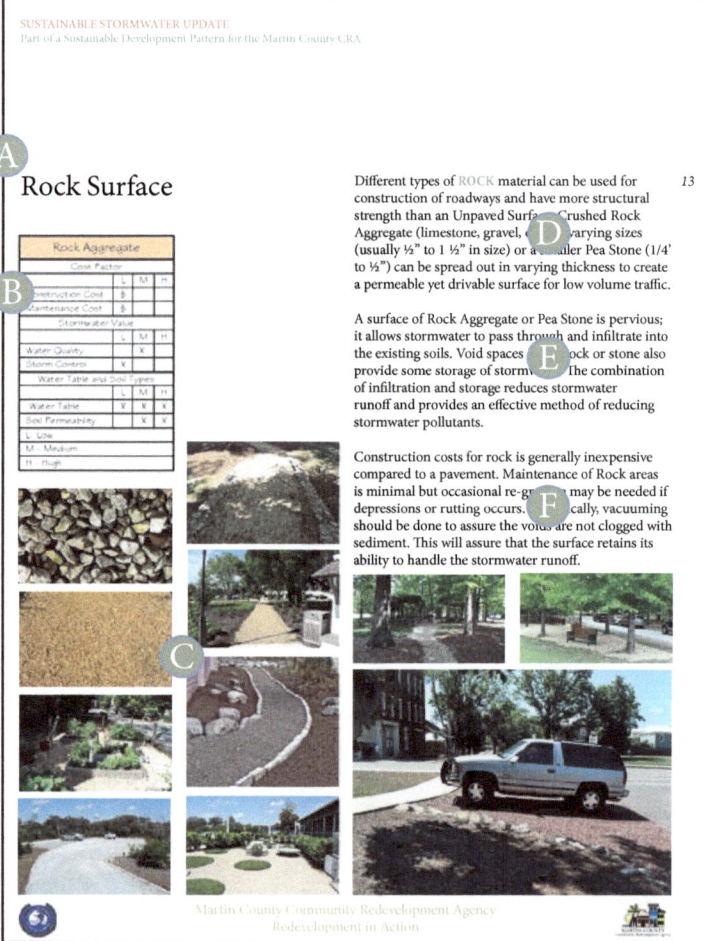

Martin County Board of County Commissioners
Your County, Your Community

Categories

 PAVEMENT — CONCRETE
 CHANNELING — INLETS
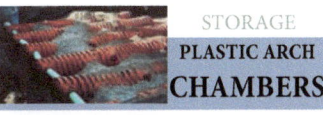 STORAGE — PLASTIC ARCH CHAMBERS
 FILTRATION — STORMWATER FILTERS

 PAVEMENT — PERVIOUS ASPHALT
 CHANNELING — GRASS SWALE
 STORAGE — CONCRETE CHAMBERS
 FILTRATION — SOAKING AREA

 PAVEMENT — PERVIOUS PAVERS
 CHANNELING — CURBS GUTTERS
 STORAGE — CONSTRUCTED PONDS
 FILTRATION — EXFILTRATION TRENCH

 PAVEMENT — PERVIOUS CONCRETE
 CHANNELING — PIPES
 STORAGE — UNDER GROUND
 FILTRATION — TREE BOX

 PAVEMENT — ASPHALT
 STORAGE — DRY RETENTION
 FILTRATION — RAIN GARDEN

 FILTRATION — BIO-RETENTION SWALES

As illustrated above, strategies have been divided into four categories: pavement, channeling, storage and filtration. These methods can be utilized independently or in combination with other stormwater management tools for maximum efficiency.

Tool Box

Pavement
Channeling
Storage
Filtration

Pavement

PAVING creates a safe, walkable, drivable surface. Paving choices vary in their ability to manage stormwater; stabilize a surface and their ease of maintenance.

Pavement Strategies

- UNPAVED SURFACES
- ROCK SURFACE
- LIME AND SHELL ROCK
- CONCRETE GRID PAVERS
- PLASTIC GRID SYSTEMS
- WOODEN WALKWAYS
- STANDARD PAVERS
- CONCRETE PAVEMENT
- ASPHALT PAVEMENT
- PERVIOUS ASPHALT PAVEMENT
- PERVIOUS CONCRETE
- PERVIOUS PAVERS
- FLEXIBLE PERVIOUS PAVEMENT

Unpaved Surfaces

The simplest form of roadway is an UNPAVED SURFACE. This can consist of existing soil material or existing soil with a grassed cover. Additional structural stabilizing material can be added to provide more strength.

Unpaved surfaces are pervious; stormwater infiltrates into the underlying soils, greatly reducing runoff. Highly permeable soils usually consist of clean sands that may not have sufficient structural strength to support heavy traffic. When stabilizing materials are added, the unpaved surface becomes less pervious because the additives are typically of a finer texture.

Construction of unpaved roadways is very inexpensive as compared to hard surfaces but they require extensive maintenance and re-grading to keep a smooth drivable surface. Grass cover can help stabilize the surface but should only be used in areas with light traffic as the grasses would not survive this volume. Surfaces without grass cover are highly susceptible to rutting and erosion during storm events.

Unpaved Surfaces			
Cost Factor			
	L	M	H
Construction Cost	$		
Maintenance Cost		$	
Stormwater Value			
	L	M	H
Water Quality		X	
Storm Control	X		
Water Table and Soil Types			
	L	M	H
Water Table	X	X	X
Soil Permeability		X	X
L - Low			
M - Medium			
H - High			

Rock Surface

Different types of ROCK material can be used for construction of roadways and have more structural strength than an Unpaved Surface. Crushed Rock Aggregate (limestone, gravel, etc.) of varying sizes (usually ½" to 1 ½" in size) or a smaller Pea Stone (1/4' to ½") can be spread out in varying thickness to create a permeable yet drivable surface for low volume traffic.

A surface of Rock Aggregate or Pea Stone is pervious; it allows stormwater to pass through and infiltrate into the existing soils. Void spaces in the rock or stone also provide some storage of stormwater. The combination of infiltration and storage reduces stormwater runoff and provides an effective method of reducing stormwater pollutants.

Construction costs for rock is generally inexpensive compared to a pavement. Maintenance of Rock areas is minimal but occasional re-grading may be needed if depressions or rutting occurs. Periodically, vacuuming should be done to assure the voids are not clogged with sediment. This will assure that the surface retains its ability to handle the stormwater runoff.

Rock Aggregate			
Cost Factor			
	L	M	H
Construction Cost	$		
Maintenance Cost	$		
Stormwater Value			
	L	M	H
Water Quality		X	
Storm Control	X		
Water Table and Soil Types			
	L	M	H
Water Table	X	X	X
Soil Permeability		X	X

L- Low
M - Medium
H - High

Lime and Shell Rock

Limerock/Shellrock			
Cost Factor			
	L	M	H
Construction Cost		$	
Maintenance Cost		$	
Stormwater Value			
	L	M	H
Water Quality	X		
Storm Control	X		
Water Table and Soil Types			
	L	M	H
Water Table	X	X	X
Soil Permeability	X	X	X
L - Low			
M - Medium			
H - High			

LIME ROCK and **SHELL ROCK** are naturally occurring in Florida and are a less expensive option for paving of lower traffic volume areas. Some may call this a "dirt road" due to the visual conditions. When used as a paving material, lime rock and shell rock are compacted to provide a hard, strong surface but may be dusty during the dry season. Normally, these materials are used as a road base with asphalt or concrete as a final surface.

Lime rock and Shell rock absorb limited moisture; therefore are basically impervious and provide little or no stormwater treatment.

Maintenance costs of Lime rock and Shell rock can be moderately expensive. Stormwater runoff can cause erosion which creates depressions and ruts in the rock surface. Higher traffic volumes will exacerbate this process. Lime rock or Shell rock surfaces should be re-graded 2 to 4 times a year and more often with higher traffic volumes. During dry conditions, dust from the Lime rock or Shell rock surface can be an air quality issue and may require regular applications of water to minimize the dust impact.

Concrete Grid Pavers

CONCRETE GRID PAVERS are open-celled concrete paver blocks. The open cells in the Concrete Grid Pavers can be filled with soil, rock, or pea stone. Concrete Grid Pavers provide the structural support of standard Concrete Pavers with the advantages of a pervious. Concrete Grid Pavers provide larger openings than pervious pavers and with a grassed surface, allow additional biological treatment and uptake of nutrients.

Concrete Grid Pavers provide stormwater storage and treatment. As with other types of pervious pavements, the effectiveness of Concrete Grid Pavers is related to the underlying soil material.

The construction cost of Concrete Grid Pavers can be high because, like other Paver systems, they are typically installed by hand; however the maintenance of Concrete Grid Pavers is minimal. With a grass cover, regular mowing is required. When rock is used in the cells, rock material will need to be replaced as needed. Concrete Grid Pavers should be inspected periodically to assure the cells remain free from silt and sediment.

Concrete Grid Pavers			
Cost Factor			
	L	M	H
Construction Cost			$
Maintenance Cost	$		
Stormwater Value			
	L	M	H
Water Quality		X	
Storm Control	X		
Water Table and Soil Types			
	L	M	H
Water Table	X	X	X
Soil Permeability		X	X

L - Low
M - Medium
H - High

STORMWATER DESIGN TOOLKIT
Martin County Community Redevelopment Agency

Plastic Grid Systems

PLASTIC GRID SYSTEMS are used to add structural stability to unpaved roadways by spreading out wheel loads. They also provide erosion control by helping to hold soils and sediments in place. The grid systems come is various shapes and are installed at or just below the finished grade. The grids or pockets are filled with soil, grass, rock or pea stone.

Plastic Grid Systems, as with other unpaved areas, provide a pervious surface that allows stormwater to infiltrate into the underlying soils some limited temporary storage of stormwater can be provided in the void spaces when rock or pea stone is used.

Initial construction cost of Plastic Grid Systems is expensive but it reduces ongoing maintenance cost as these grids reduce erosion and rutting. For grassed areas, periodic mowing is generally all that is required for maintenance.

Plastic Grid Systems			
Cost Factor			
	L	M	H
Construction Cost		$	
Maintenance Cost	$		
Stormwater Value			
	L	M	H
Water Quality		X	
Storm Control	X		
Water Table and Soil Types			
	L	M	H
Water Table	X	X	X
Soil Permeability		X	X
L- Low			
M - Medium			
H - High			

Credit: R.A.Smith National

Martin County Board of County Commissioners
Your County, Your Community

Wooden Walkways

WOODEN WALKWAYS can be a pervious alternative to paved sidewalks. Wooden Walks are generally used for ramps or boardwalks above the ground surface but can be imbedded directly in the soil.

Wooden Walkways are pervious and allow stormwater to infiltrate into the soil. Storage can be enhanced by creating a depression area under the Walkways to temporarily hold some stormwater runoff. Wooden Walkways can also be an effective treatment for stormwater. Its efficiency is determined by the permeability of the underlying soil.

Construction of Wooden Walkways can be more expensive than concrete. The cost varies with the type of wood used. Wooden Walkways need to be painted, stained or sealed at installation and periodically as needed. They should have regular inspections to assure the wood continues to be structurally sound. As an alternative, there are plastic, composite and re-cycled products. Although they are more expensive they require little or no maintenance and have a longer lifespan.

Wood Walks			
Cost Factor			
	L	M	H
Construction Cost		$	
Maintenance Cost		$	
Stormwater Value			
	L	M	H
Water Quality		X	
Storm Control	X		
Water Table and Soil Types			
	L	M	H
Water Table	X	X	X
Soil Permeability		X	X
L - Low			
M - Medium			
H - High			

Standard Pavers

Standard Pavers			
Cost Factor			
	L	M	H
Construction Cost			$
Maintenance Cost	$		
Stormwater Value			
	L	M	H
Water Quality	X		
Storm Control	X		
Water Table and Soil Types			
	L	M	H
Water Table	X	X	X
Soil Permeability	X	X	X
L - Low			
M - Medium			
H - High			

STANDARD PAVERS are a decorative alternative for roadway surface of today. Various materials may be used such as concrete, brick or stone and come in a variety of shapes, sizes, colors and textures. Typically, Pavers are constructed on existing soil and rock base similar to the layers under Asphalt Pavement. The Pavers are usually set in a thin bed of sand to allow for leveling of the surface.

Paver material is impervious however its installation determines its ability to handle stormwater runoff. Typical installation is to fill the space between each paver with cement mortar or a thin sheet of sand which hardens. This option is impervious. However, a joint profile paver can be used and installed with washed stone aggregate (which is larger than a grain of sand but smaller than a pebble). This installation and paver allows surface water to infiltrate into the pavement stormwater sub layers creating some stormwater treatment.

Labor costs of pavers can be expensive as they are normally placed in position by hand. Periodic maintenance is generally limited to sweeping to remove sediments and debris so maintenance costs are relatively low. They also are easy to repair as pavers can be individually replaced as they are damaged. Overall replacement cost is much lower than other forms of paving material such as asphalt or concrete.

Concrete Pavement

CONCRETE PAVEMENT is another form of roadway surface. Concrete Pavement consists of sand and rock materials mixed with cement, creating a firm pavement surface. Concrete Pavement can be constructed on compacted existing soil but can be fortified when used with a lime rock or shell rock base. Concrete Pavement has added benefits. It can be stamped with patterns or be colorized for visual affect.

As with other types of Standard Pavement systems, Concrete Pavement is impervious and creates stormwater runoff; therefore it provides no stormwater quality treatment.

Concrete Pavement is usually more expensive than Asphalt Pavement; however its lifespan can be from 15 to 20 years. As with Asphalt Pavement, regular maintenance costs are low but periodic sweeping is recommended to keep materials out of stormwater drainage systems.

Concrete Pavement			
Cost Factor			
	L	M	H
Construction Cost			$
Maintenance Cost	$		
Stormwater Value			
	L	M	H
Water Quality	X		
Storm Control	X		
Water Table and Soil Types			
	L	M	H
Water Table	X	X	X
Soil Permeability	X	X	X
L - Low			
M - Medium			
H - High			

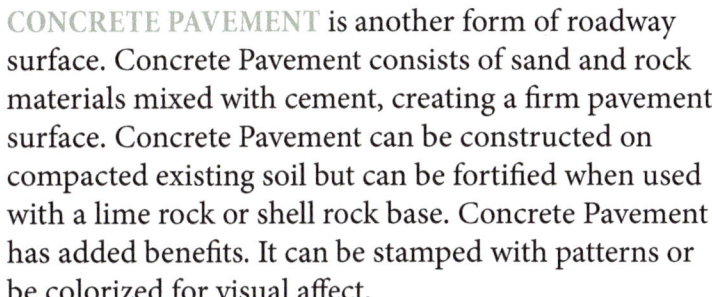

Asphalt Pavement

ASPHALT PAVEMENT is generally the most common paving material. Asphalt Pavement consists of a liquid bituminous material that acts as a binder for sand and rock. Asphalt is constructed over existing soils enhanced with additional material such as a lime rock or crushed shell base to provide structural support.

Asphalt Pavement is impervious (does not allow water to pass through) creating stormwater runoff; therefore Asphalt Pavement surfaces provide no stormwater quality treatment.

Asphalt Pavement is usually the least expensive roadway surface. If properly constructed, Asphalt Pavement systems last for 12 to 15 years. Regular maintenance is limited but regular cleaning and sweeping is recommended to prevent trash and debris from getting into the stormwater drainage system.

Asphalt Pavement			
Cost Factor			
	L	M	H
Construction Cost		$	
Maintenance Cost	$		
Stormwater Value			
	L	M	H
Water Quality	X		
Storm Control	X		
Water Table and Soil Types			
	L	M	H
Water Table	X	X	X
Soil Permeability	X	X	X
L - Low			
M - Medium			
H - High			

Pervious Asphalt Pavement

PERVIOUS ASPHALT PAVEMENTS utilize a mixture of liquid asphalt materials mixed with fine sand and larger rock to create a pavement surface with open void spaces. This surface allows stormwater to flow through and infiltrate into the underlying soil layers. Due to the different rock material used, Pervious Asphalt has a rougher surface texture than the smooth texture of traditional asphalt pavement.

Pervious Asphalt is pervious and allows stormwater to pass through and provides for storage in the void space between the effectiveness of this pavement is dependent upon the permeability of the underlying soil.

The material cost of Pervious Asphalt Pavement is slightly higher than standard asphalt pavement but the lower costs of design, permitting and installation significantly offsets this. Pervious Asphalt Pavement is best suited for locations with lower volume traffic such as residential areas. This surface should be routinely vacuumed to maintain its drainage potential.

Pervious Asphalt			
Cost Factor			
	L	M	H
Construction Cost			$
Maintenance Cost		$	
Stormwater Value			
	L	M	H
Water Quality			X
Storm Control		X	
Water Table and Soil Types			
	L	M	H
Water Table	X	X	X
Soil Permeability		X	X
L - Low			
M - Medium			
H - High			

Pervious Concrete

Pervious Concrete			
Cost Factor			
	L	M	H
Construction Cost			$
Maintenance Cost		$	
Stormwater Value			
	L	M	H
Water Quality			X
Storm Control		X	
Water Table and Soil Types			
	L	M	H
Water Table	X	X	X
Soil Permeability		X	X
L - Low			
M - Medium			
H - High			

PERVIOUS CONCRETE PAVEMENT uses a mixture of cement and rock material of various sizes that creates void spaces to allow water to flow through. Pervious Concrete Pavement generally has a rougher texture than traditional concrete and is often constructed with a rock bed underneath to allow for additional stormwater storage.

Pervious Concrete provides drainage benefits which, as with other infiltration methods, its effectiveness is dependent upon the underlying soils.

The construction cost of Pervious Concrete Pavement is higher than standard concrete pavement. The use of Pervious Concrete Pavement can greatly reduce the volume of stormwater runoff. Pervious Concrete Pavement should be vacuumed once or twice a year, to remove silts and sediment that can clog the pavement pores. Pervious Concrete is generally limited to low volume areas.

Pervious Pavers

PERVIOUS PAVERS are constructed with a gap or space that allows stormwater to infiltrate into the underlying soil. Pervious Pavers come in a variety of materials, colors and configurations. Pervious Pavers can be made of concrete, brick or natural stone.

Pervious Pavers provide stormwater treatment with an attractive appearance. This space between paver units is usually filled with highly permeable sands or small stones to allow for rapid infiltration. As with other pervious pavements, its effectiveness is related to the underlying soil. The use of Pervious Pavers reduces the volume of stormwater runoff and therefore can be an effective way of reducing pollutants.

The construction cost of Pervious Pavers is slightly higher than standard Pavers. Pervious Pavers should be swept once or twice a year and should be vacuumed to remove silts and sediment.

Pervious Pavers			
Cost Factor			
	L	M	H
Construction Cost			$
Maintenance Cost		$	
Stormwater Value			
	L	M	H
Water Quality			X
Storm Control		X	
Water Table and Soil Types			
	L	M	H
Water Table	X	X	X
Soil Permeability		X	X
L - Low			
M - Medium			
H - High			

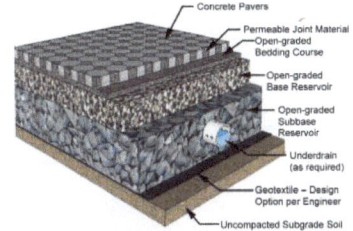

Flexible Pervious Pavement

FLEXIBLE PERVIOUS PAVEMENT consists of flexible particles of varying sizes (often rubber chips) with an adhesive binder agent that creates open voids in the pavement that allows stormwater to flow through the pavement section to the underlying soil material. Re- cycled tires are often used as part of the flexible material.

Flexible Pervious Pavement provides the benefit of reducing stormwater runoff and off-site discharges of stormwater pollutants. It is rarely used for roadways. This type of pavement comes with a selection of colors that can provide a more pleasing visual affect.

Construction costs of Flexible Pervious Pavement are higher than traditional pavements such as concrete or asphalt. Vacuuming once or twice a year is required in order to keep the void spaces open for flow, and infiltration.

Flexible Pervious Pave			
Cost Factor			
	L	M	H
Construction Cost			$
Maintenance Cost		$	
Stormwater Value			
	L	M	H
Water Quality			X
Storm Control		X	
Water Table and Soil Types			
	L	M	H
Water Table	X	X	
Soil Permeability		X	X
L - Low			
M - Medium			
H - High			

SUSTAINABLE STORMWATER UPDATE

Case Study

Palm City Pervious

34th Terrace and Noble Avenue • Old Palm City

Innovation and sustainability does not always mean more cost. In the Old Palm City Community Redevelopment Area, staff from various Martin County Departments collaborated to improve an unpaved residential street. Pervious Asphalt was identified as the best solution for 1,700 linear feet of residential streets with lower traffic volumes. Although the initial material cost is slightly higher than standard asphalt paving, the total project cost is much lower, and the project timeline was dramatically reduced.

In this case study, Martin County Staff managed the entire concept and implementation of this project which was priced using an existing approved annual paving contractor. The combined linear footage total for 34th Terrace and Noble Avenue is 1,700 linear feet or .32 miles. The total project cost was $38 a linear foot for a total cost of $66,850. This project was conceived and implanted in a matter of weeks because of the affordability, ease of installation, and stormwater benefits, found with pervious asphalt.

In Martin County, the average cost of a standard asphalt roadway with typical stormwater management practices is one million dollars a linear mile. This figure includes fees for engineering services, permitting, installation of infrastructure such as pipes, gutters and curbing and cost of sub-grade materials. In contrast, innovative material choices like pervious asphalt can be installed directly over a dirt road without the need for exhaustive engineering drawings, state permits and other underground and off-site infrastructure for a cost of $320,000.00 a linear mile. Innovative material choices like pervious asphalt demonstrates an 80% cost savings to Martin County residents compared to conventional practices.

Project Type
Infrastructure
Innovative Stormwater Management

Partnerships
Martin County Engineering Department
Martin County Board of County Commissioners
Martin County Community Redevelopment Agency

Funding
Tax Increment Financing (TIF)

Project Size
1,700 Linear Feet or .32 Miles

Year Completed
Summer 2011

ABOVE: Neighborhood Streets Prior Condition
RIGHT: Neighborhood Streets After Pervious Paving

Martin County Community Redevelopment Agency
Redevelopment in Action

Channeling

CHANNELING is a method to collect and direct stormwater to specific locations for flood prevention and/or water quality treatment.

Channeling Strategies

- **GRASSED SWALE**
- **DITCHES**
- **CANALS**
- **CURBS AND GUTTERS**
- **STORMWATER INLETS**
- **STORMWATER PIPING**

Grassed Swale

A **SWALE** is a shallow water channel normally with a grass cover. Swales collect and direct stormwater flows on sides of roadways and are an alternative to curbs, gutters, inlets and piping.

Swales generally provide minimal stormwater storage and limited water quality treatment. Swales provide more water treatment than curbs and gutter because they are pervious and allow filtration.

Grassed Swales are inexpensive to construct. Maintenance is limited to mowing and periodically removing debris. Occasional maintenance of slopes is needed to assure their efficiency.

Swales			
Cost Factor			
	L	M	H
Construction Cost	$		
Maintenance Cost	$		
Stormwater Value			
	L	M	H
Water Quality	X		
Storm Control	X		
Water Table and Soil Types			
	L	M	H
Water Table	X	X	
Soil Permeability		X	X
L - Low			
M - Medium			
H - High			

Ditches

DITCHES are similar to Swales but are deeper with steeper side slopes. Ditches can be either dry or wet depending on the depth and water table. Large Ditches that permanently hold water are designated as Canals. Ditches are usually grassed. The steep slopes may require erosion control fabrics or rock rip-rap for stabilization.

Ditches provide limited temporary stormwater storage and some water quality treatment.

Ditches are inexpensive to construct and maintain. Maintenance involves mowing and maintaining vegetation. Grass and other growth must be kept to a minimum to avoid contributing to flooding.

Ditches			
Cost Factor			
	L	M	H
Construction Cost	$		
Maintenance Cost	$		
Stormwater Value			
	L	M	H
Water Quality	X		
Storm Control	X		
Water Table and Soil Types			
	L	M	H
Water Table		X	X
Soil Permeability		X	X
L - Low			
M - Medium			
H - High			

Canals

Canals			
Cost Factor			
	L	M	H
Construction Cost			$
Maintenance Cost		$	
Stormwater Value			
	L	M	H
Water Quality		X	
Storm Control			X
Water Table and Soil Types			
	L	M	H
Water Table		X	X
Soil Permeability		X	X
L- Low			
M - Medium			
H - High			

CANALS are large surface stormwater facilities that are larger than Ditch systems and are often owned and maintained by municipalities. In addition there are a number of canal systems within large privately owned agricultural properties. Typically, Canal systems have relatively steep side slopes and maintain several feet of water even during low water conditions. Canal banks often are reinforced with erosion control fabrics or rock rip-rap.

Canals can be a relatively effective stormwater quality treatment method.

Construction costs and excavation can be high, but on-going costs for maintenance are moderate to high. Maintenance is generally limited to periodic inspections with mowing and maintaining native vegetation. In addition, it may be necessary to remove nuisance aquatic vegetation growth that can affect the flow capacity of the canal system.

Curbs and Gutters

CURBS AND GUTTERS are typically constructed at the edge of pavement on roadways and parking areas to collect stormwater runoff and direct it into Stormwater Inlets or other drainage facilities. Curbs are from 4 inches to 6 inches high, and can be constructed alone or combined as an integrated Curb and Gutter system. Gutters can be constructed with or without a Curb section. Curbs and Gutters are typically constructed of concrete but can be of other materials including asphalt, concrete or brick blocks or pavers, and natural stone. Curbs and Gutters are most often constructed in urban areas where limited right-of-way widths prevent the construction of grassed swales.

Curbs and Gutters are designed to collect and carry stormwater and provide no storage or water quality treatment.

Although more expensive than swales, installation costs for Curbs and Gutters can be relatively low to moderate. Maintenance requirements are minimal and may consist of sweeping and removal of debris and sediments.

Curbs and Gutters			
Cost Factor			
	L	M	H
Construction Cost		$	
Maintenance Cost	$		
Stormwater Value			
	L	M	H
Water Quality	X		
Storm Control	X		
Water Table and Soil Types			
	L	M	H
Water Table	X	X	X
Soil Permeability	X	X	X
L - Low			
M - Medium			
H - High			

Stormwater Inlets

Stormwater Inlets			
Cost Factor			
	L	M	H
Construction Cost		$	
Maintenance Cost		$	
Stormwater Value			
	L	M	H
Water Quality	X		
Storm Control	X		
Water Table and Soil Types			
	L	M	H
Water Table	X	X	X
Soil Permeability	X	X	X
L - Low			
M - Medium			
H - High			

STORMWATER INLETS are normally made of concrete (also known as catch basins) and are designed to collect stormwater. They function as part of an interconnected system of underground piping that discharges to either a stormwater retention/detention facility or to a stormwater outfall to an existing water body. They also have cast iron or steel grates of varying sizes and shapes. Standard Stormwater inlets are designed to collect stormwater but do not provide storage or treatment.

Construction cost for Stormwater Inlets is generally moderate but can be high depending on the size and type of structure needed. Maintenance consists of inspections and removal of accumulated trash, debris and sediments from both the surface grate and inside the Inlet structure and is often done with a vacuum truck system. Maintenance costs can be low to moderate depending on the number of Inlets and the potential for debris (leaves, sand and silt) that may come from the contributing site area.

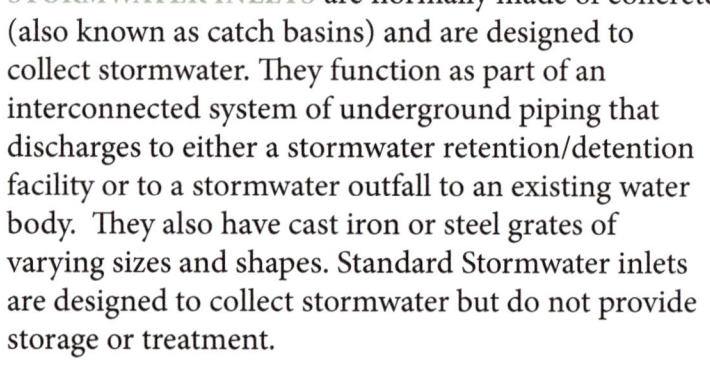

Stormwater Piping

Stormwater Piping			
Cost Factor			
	L	M	H
Construction Cost	$	$	$
Maintenance Cost	$		
Stormwater Value			
	L	M	H
Water Quality	X		
Storm Control	X		
Water Table and Soil Types			
	L	M	H
Water Table	X	X	X
Soil Permeability	X	X	X
L - Low			
M - Medium			
H - High			

STORMWATER PIPES are either concrete, plastic or metal from 4 to 6 inch diameters up to 72" diameter for larger drainage areas.

Stormwater Piping carries stormwater downstream and is not for stormwater treatment, storage or control.

Costs for installation for Stormwater Piping can vary substantially depending on the size and material of the pipe and the depth of excavation needed for specific projects. Maintenance is usually limited to an annual or semi-annual inspection of the piping system and removal of any accumulated trash, debris and sediments. This is often done with a vacuum truck and maintenance costs are low to moderate depending on the length of pipe to be cleaned.

Storage

STORAGE

Stormwater can be stored by retaining it for a period of time to allow soils to naturally filter the water. It can also be stored and re-used for irrigation and other non-potable uses.

Storage Strategies

- **DRY RETENTION/DETENTION AREAS**
- **UNDER DRAINS**
- **UNDER-GROUND STORAGE**
- **CONSTRUCTED PONDS AND LAKES (WET RETENTION/ DETENTION)**
- **LINED LAKES, PONDS AND POOLS**
- **POOLS AND FOUNTAINS**
- **STORMWATER HARVESTING**
- **ROOF-TOP STORAGE**
- **CISTERNS**
- **RAIN BARRELS**
- **PLASTIC CELLULAR CHAMBERS**
- **CONCRETE CHAMBERS**
- **PLASTIC ARCH CHAMBERS**

Dry Retention/Detention Areas

Dry Ret./Det. Areas			
Cost Factor			
	L	M	H
Construction Cost	$		
Maintenance Cost		$	
Stormwater Value			
	L	M	H
Water Quality			X
Storm Control			X
Water Table and Soil Types			
	L	M	H
Water Table	X	X	
Soil Permeability		X	X
L - Low			
M - Medium			
H - High			

DRY RETENTION/DETENTION AREAS are man-made recessed areas of land that are used as a method of stormwater control, and treatment. These types of stormwater storage areas are designed to temporarily hold stormwater runoff and either allows the water to infiltrate into the soil with no discharge (retention) or a combination of infiltration and discharge at a controlled rate through an outfall structure (detention). Since Dry Retention/Detention Areas rely on infiltration, these types of stormwater systems work best in existing soils with moderate to high permeability rates.

The bottom of a Dry Retention/Detention Area should be a minimum of one (1) foot above the expected seasonal high water table to be considered a dry system. Because of this requirement, Dry Retention/Detention Areas located on properties with high water tables must be shallower and therefore require a larger area for a given volume of stormwater runoff.

Construction cost of a Dry Retention/Detention Area is generally inexpensive but can take up a significant amount of surface area to meet stormwater treatment and storage requirements. Long-term maintenance consists of periodic mowing of grass and landscape maintenance. An irrigation system may be necessary during dry seasons in order to maintain grass and plant material.

SUSTAINABLE STORMWATER UPDATE

Under Drains

Dry with Underdrains			
Cost Factor			
	L	M	H
Construction Cost		$	
Maintenance Cost	$		
Stormwater Value			
	L	M	H
Water Quality		X	
Storm Control		X	
Water Table and Soil Types			
	L	M	H
Water Table	X	X	
Soil Permeability		X	X
L - Low			
M - Medium			
H - High			

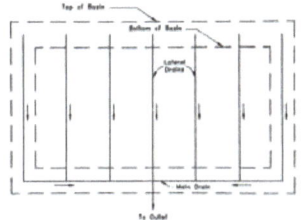

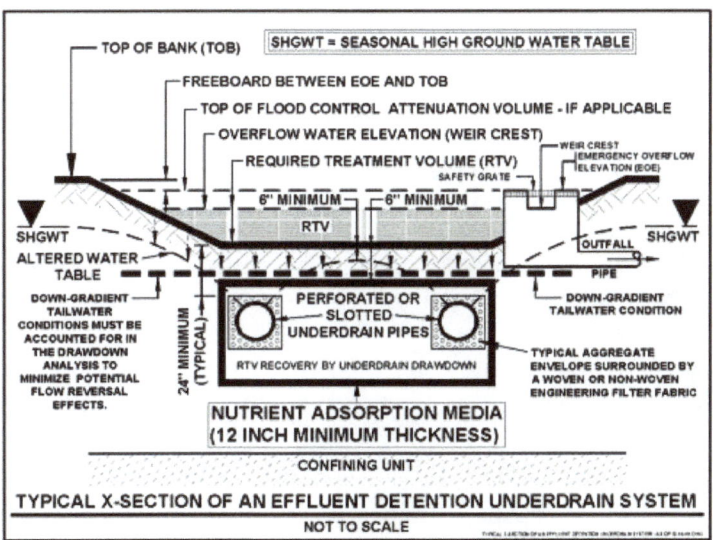

UNDER DRAINS are best when the existing soils in the area of a Dry Retention/Detention Areas have a low permeability, an under drain is utilized to control the water level.

Under drains are constructed of perforated piping (PVC, HDPE, or Metal) in a linear trench pattern below the bottom of the Dry Retention/Detention Areas. The perforated pipes are placed in a trench of rock material that is wrapped in filter cloth. This helps maintain the permeability of the system by preventing fine sands and silt from clogging the voids in the rock and plugging the perforations in the pipe.

Under drains provide some water quality treatment because they filter stormwater prior to collection in the under drain piping; however the reduction of pollutants and nutrients is somewhat limited.

Adding the under drain system will increase the cost of a typical Dry Retention/Detention Area. Long-term maintenance costs of retention/detention areas are the same with or without the under drain but under drains should be monitored during and after a storm. Over time, silts and sediments may accumulate and cause the under drain system to fail. The under drain should be cleaned, excavated or replaced as needed.

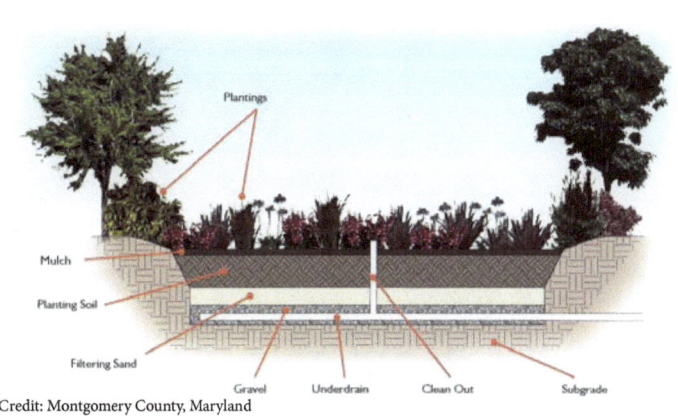

Credit: Montgomery County, Maryland

Martin County Community Redevelopment Agency
Redevelopment in Action

Under-Ground Storage

Many stormwater BMP's are designed to direct stormwater underground and allow the surface area to be used for other facilities such as parking, roadways and landscaping. As an alternative, stormwater can also be directed to an underground storage area beneath buildings and roadways for storage and re-use or for infiltration into the underlying soils. Underground Storage may consist of a vault system, typically of concrete boxes, that can store water for irrigation or other non-potable use such as toilet flushes. Another type of Underground Storage system has an open bottom to allow infiltration. Stormwater is collected and discharged to the area below the vault and allowed to infiltrate into the surrounding soils. Because this area is under the vault and permanently in the shade, usually with this type of storage the surface consists of rock material.

By reducing the volume of stormwater runoff that is discharged off site, an Underground Storage system can be a highly effective water treatment process.

Construction cost is high because the roadway surface or the first floor of a building must be designed to be structurally sound to support the underground storage area. In addition, this type of system may require special water-proofing considerations on the concrete boxes or the underside of the building floor. Maintenance costs are moderate and consist of inspections of the undergrounds storage area and removal of any accumulated leaves, trash and debris.

Under Building Storage			
Cost Factor			
	L	M	H
Construction Cost			$
Maintenance Cost		$	
Stormwater Value			
	L	M	H
Water Quality			X
Storm Control		X	
Water Table and Soil Types			
	L	M	H
Water Table	X	X	
Soil Permeability		X	X
L - Low			
M - Medium			
H - High			

Credit: Catalano Construction, Inc.

Credit: Vertex Design Group

Constructed Ponds and Lakes (Wet Retention/Detention)

Constructed Pond/Lake			
Cost Factor			
	L	M	H
Construction Cost		$	
Maintenance Cost		$	
Stormwater Value			
	L	M	H
Water Quality			X
Storm Control			X
Water Table and Soil Types			
	L	M	H
Water Table		X	X
Soil Permeability	X	X	X
L - Low			
M - Medium			
H - High			

CONSTRUCTED PONDS AND LAKES are permanent wet systems designed for stormwater treatment and flood protection. They have an added benefit of providing aquatic habitats and esthetic features as well as opportunities for water-related recreation and use as an irrigation source. Ponds and Lakes are generally restricted to certain minimum size requirements in order to provide sufficient treatment and stormwater storage.

Ponds and Lakes are highly effective stormwater treatment systems. Their effectiveness is related to the permanent volume of water in the Pond or Lake and overall time it takes for stormwater to pass through and out of the Pond or Lake System.

Cost for excavation, grading, landscape and shoreline planting can be moderate to high. Ponds and Lakes are often constructed on sites where fill is needed so the initial costs can be offset by a reduction in the costs of off-site fill material. Long-term maintenance costs are moderate and include maintenance of landscape and shoreline plant material. Stormwater facilities at the Pond or Lake including pipes and structures should be inspected periodically. Sand, silt, debris or trash should be removed as needed to assure efficiency.

Lined Lakes, Ponds and Pools

Lined Lake, Pond, Pool			
Cost Factor			
	L	M	H
Construction Cost		$	
Maintenance Cost		$	
Stormwater Value			
	L	M	H
Water Quality			X
Storm Control			X
Water Table and Soil Types			
	L	M	H
Water Table	X	X	X
Soil Permeability	X	X	X
L - Low			
M - Medium			
H - High			

LINED LAKES, PONDS AND SMALL POOLS are constructed with an impervious liner to prevent the infiltration and loss of usable water. This may be done for esthetic purposes to create a water feature at an elevation above normal water table elevations or may be associated with irrigation systems as a way to maximize the storage of irrigation quality water. The liner material is typically a specialized clay material or a plastic impervious mat or a combination of both that is constructed below the bottom of the Pond. Lined Ponds are often used as storage facilities for treated irrigation quality wastewater also called re-use or re-claimed water. Lined Ponds must have an overflow structure to allow for discharge once the maximum Pond storage capacity is reached.

Depending on the total retained water volume, a Lined Lake, Pond or Pool can be a very effective water quality treatment method, particularly if the stormwater volumes are re-used for irrigation.

Adding the liner increases the construction cost of a pond; but maintenance costs are similar for those of a standard Pond or Lake System. The water elevations must be monitored to identify problems with the impervious liner system. Cost to replace the liner can be substantial depending on the size of the Lake, Pond or Pool.

Pools and Fountains

Decorative **POOLS AND FOUNTAINS** an aesthetic storage and treatment option that can be included as part of the landscaping features of development projects.

Pools and Fountains can contribute somewhat to the stormwater treatment process. The effectiveness of stormwater treatment with Pools and Fountains is relatively small and is related to the overall volume of water that can be stored and the ability of any aquatic vegetation in the system to absorb nutrients and stormwater pollutants. Screens, filters and sedimentation are used with Pools and Fountains to remove debris. Pools and Fountains normally have a separate pump and filter system with an overflow drain.

Costs for Pools and Fountains can vary depending upon how elaborate the systems are and whether they include sculptures and other artistic features. Pools and Fountains have moderate to extensive maintenance and includes cleaning filter systems, maintaining the pump systems and periodic removal of dirt, leaves and debris that may have accumulated.

Pools and Fountains			
Cost Factor			
	L	M	H
Construction Cost		$	
Maintenance Cost	$		
Stormwater Value			
	L	M	H
Water Quality	X		
Storm Control	X		
Water Table and Soil Types			
	L	M	H
Water Table	X	X	X
Soil Permeability	X	X	X
L - Low			
M - Medium			
H - High			

Stormwater Harvesting

Stormwater Harvesting			
Cost Factor			
	L	M	H
Construction Cost		$	
Maintenance Cost	$		
Stormwater Value			
	L	M	H
Water Quality		X	
Storm Control	L		
Water Table and Soil Types			
	L	M	H
Water Table	X	X	X
Soil Permeability		X	X
L - Low			
M - Medium			
H - High			

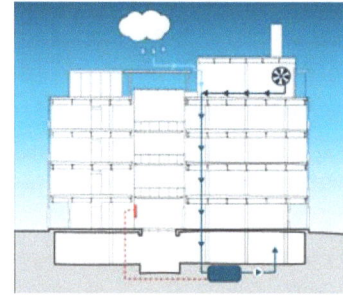

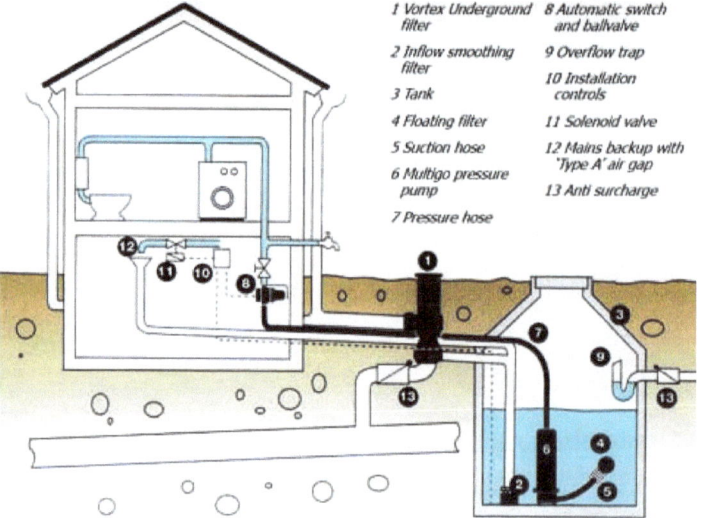

STORMWATER HARVESTING is the reuse of stormwater runoff after it has been collected in some type of retention system. The use of Cisterns, Rain Barrels and other types of Stormwater Reuse system are examples of Stormwater Harvesting. Another method of Stormwater Harvesting is the draw down of stormwater runoff from ponds or lakes. Lake water levels will rise above their normal water level during storm events as runoff is collected and discharged into the lake. This retained volume can be used for irrigation or other non-potable water uses once the storm event is over.

Essentially, this type of Stormwater Harvesting system consists of a horizontal perforated well casing constructed along the side bank of a pond or lake. The stormwater runoff temporarily stored in the lake is filtered and treated as it is drawn through the soil material in the lake bank and is pumped out for use as landscape irrigation. The horizontal well system is designed to only draw down the lake to the normal water level and does not draw from deeper into the water table.

Since Stormwater Harvesting involves direct reuse of retained stormwater runoff, this method is a very effective water quality treatment method for reduction of pollutants and nutrients.

Initial construction costs are moderate to expensive depending on the size of the system. Regular on-going maintenance costs are relatively low and are similar to other irrigation well and pump systems. Since this type of system is dependent on stormwater runoff as a water source, it could be necessary to have a supplemental irrigation source during dry seasons.

SUSTAINABLE STORMWATER UPDATE

Cisterns

Cisterns			
Cost Factor			
	L	M	H
Construction Cost		$	
Maintenance Cost	$		
Stormwater Value			
	L	M	H
Water Quality		X	
Storm Control		X	
Water Table and Soil Types			
	L	M	H
Water Table	X	X	X
Soil Permeability	X	X	X
L - Low			
M - Medium			
H - High			

CISTERNS have been used for hundreds of years and are a method of collecting, storing and re-using stormwater runoff for multiple purposes. Cisterns use large above-ground or underground storage tanks made of concrete, fiberglass, metal or plastic materials. Historically, Cisterns have been used to collect and store rainfall for drinking water and are still used for this purpose today. Generally, Cistern systems in Florida are used to store water for landscape irrigation. Cisterns are also often used for storing roof runoff to irrigate Green Roofs.

Cistern systems provide stormwater quality treatment and storm storage. The effectiveness is limited based upon the total stormwater runoff volume stored and re-used.

The construction cost for Cisterns depends upon the size and material used. Pre-fabricated fiberglass or plastic tanks are relatively inexpensive while pre-cast concrete tanks are more costly. Maintenance cost is relatively low. Periodic inspections are required for Cisterns to make sure that the systems are water-tight and functioning properly.

Rain Barrels

Rain Barrels			
Cost Factor			
	L	M	H
Construction Cost	$		
Maintenance Cost	$		
Stormwater Value			
	L	M	H
Water Quality	X		
Storm Control	X		
Water Table and Soil Types			
	L	M	H
Water Table	X	X	X
Soil Permeability	X	X	X
L - Low			
M - Medium			
H - High			

RAIN BARRELS are a type of Cistern consisting of relatively small storage tanks that are connected to roof downspouts and collect rain runoff for re-use. The stored rainwater is normally used for landscape irrigation or other non-potable water uses. Normally Rain Barrels are made of plastic or fiberglass. Manufactured Rain Barrels come in a variety of shapes and colors and often come complete with a cover and/or screen, overflow piping and hose connections. Rain Barrels can also be home-made using readily available products and materials.

By storing and re-using rainfall, Rain Barrels provide an effective small scale method for reducing pollutant discharges in stormwater runoff. Since the stored water volume is limited, Rain Barrels are usually used on residential sites.

Installation cost for Rain Barrels is relatively low. Minimal maintenance is required for Rain Barrels and includes periodic inspections to assure the system is water-tight and functioning, and debris should be removed from the screen and tank as needed. As an additional precaution, a downspout filter can be installed to keep out leaves and other debris.

Plastic Cellular Chambers

Plastic Cellular Chambers			
Cost Factor			
	L	M	H
Construction Cost			$
Maintenance Cost		$	
Stormwater Value			
	L	M	H
Water Quality			X
Storm Control		X	
Water Table and Soil Types			
	L	M	H
Water Table	X	X	
Soil Permeability		X	X
L - Low			
M - Medium			
H - High			

PLASTIC CELLULAR EXFILTRATION CHAMBERS are manufactured plastic cubic structures that can be inter-connected to form various shapes and sizes for underground storage and exfiltration of stormwater. These structures have open spaces on the top, bottom and sides to allow water to soak into the surrounding soils. Cellular Chambers can be located under parking areas, driveways and roadways. The strength of the Cellular units is in the top, bottom, sides and internal supports of each unit thereby providing as much as 90% of the total volume for storage of stormwater. Cellular Chambers are normally placed on a bed of rock with the entire structure wrapped with filter cloth to prevent sand, silt and sediment from entering the system. Underground exfiltration systems work best in highly permeable soils with a deep depth to the water table.

Installation of Plastic Cellular Exfiltration systems will reduce overall stormwater discharges thereby providing an effective water quality treatment method for reducing stormwater pollutants.

Construction costs of Cellular Chamber are relatively expensive but this expense is offset by the increase in developable land. Maintenance costs are moderate and include inspections of the Chamber system and vacuuming of debris on a semi-annual basis.

Concrete Chambers

CONCRETE EXFILTRATION CHAMBERS consist of pre-cast or cast-in-place concrete chambers of various sizes that can be used for storing stormwater. As with other exfiltration systems, Concrete Chambers function best when the existing soils exhibit high permeability, and storage volumes can be maximized with deep water tables.

Exfiltration Chambers use retention and infiltration as an effective means of water treatment thereby substantially reducing off-site discharges of pollutant-laden stormwater.

Concrete Exfiltration Chambers can be expensive to construct but this cost can be offset by having more developable area. The area above the Chambers can be used for parking, driveways, roadways and other uses. Access ports, clean outs or manholes are typically installed with Concrete Chambers for inspection and maintenance. Maintenance cost is moderate and includes periodic inspections and to vacuum any accumulated silt or sediments.

Concrete Chambers			
Cost Factor			
	L	M	H
Construction Cost			$
Maintenance Cost		$	
Stormwater Value			
	L	M	H
Water Quality			X
Storm Control		X	
Water Table and Soil Types			
	L	M	H
Water Table	X	X	
Soil Permeability		X	X
L - Low			
M - Medium			
H - High			

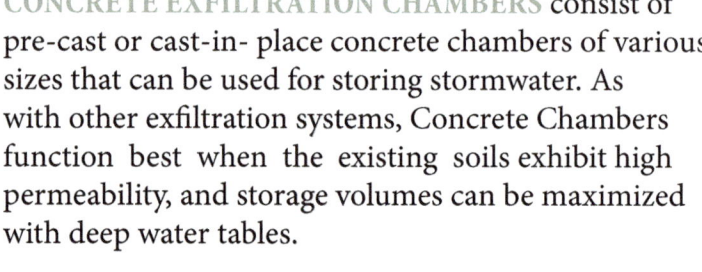

Plastic Arch Chambers

Plastic Arch Chambers			
Cost Factor			
	L	M	H
Construction Cost			$
Maintenance Cost		$	
Stormwater Value			
	L	M	H
Water Quality			X
Storm Control		X	
Water Table and Soil Types			
	L	M	H
Water Table	X	X	
Soil Permeability		X	X
L - Low			
M - Medium			
H - High			

PLASTIC ARCH EXFILTRATION CHAMBERS are similar to standard Exfiltration Trenches but utilize corrugated Arch Chambers with open bottoms rather than perforated pipe. The corrugations provide structural strength and the Chambers are of various depths and storage capacities. As with Exfiltration Trenches, the Arch structures are usually enclosed in a rock bed surrounded with a filter cloth. Surface inlets direct stormwater to the underground Arch Chambers where the stored water soaks into the surrounding soil. Plastic Arch Chambers are best suited in soils with a high permeability and with a deep depth to the water table.

Using Plastic Arch Exfiltration Chambers is an effective stormwater treatment method. By storing and filtering stormwater, overall runoff volumes are reduced and therefore discharges of nutrients and other stormwater pollutants are also reduced.

Plastic Arch Chambers are expensive to construct but these costs are often offset by the available use of the surface area above the Chambers. Regularly scheduled maintenance is limited to visual inspections with cleaning and/or vacuuming as necessary on an annual or semi-annual basis.

Filtration

FILTRATION is a method to direct water to a specific location to allow natural soils and plants and other man-made filtering systems to absorb pollutants from the stormwater. The efficiency of the process depends upon the type of soil, plant or device utilized in the process.

Filtration Strategies

- **NATURAL PONDS AND LAKES**
- **NATURAL WETLANDS**
- **CONSTRUCTED WETLANDS**
- **BIO-RETENTION SWALES**
- **RAIN GARDENS**
- **LITTORAL ZONE PLANTING**
- **TREE BOX FILTERS**
- **ALGAE MAT**
- **GREEN ROOF**
- **VEGETATED FILTER STRIPS**
- **EXFILTRATION TRENCH**
- **SAND FILTERS**
- **ROCK SOAKING AREA**
- **FLOATING VEGETATED MATS**
- **VEGETATED WALL**
- **STORMWATER FILTERS**
- **INLET SCREENS, BAFFLES, SUMPS**
- **BAFFLE BOXES**
- **VORTEX SEPARATORS**
- **CHEMICAL (ALUM) TREATMENT**

Natural Ponds and Lakes

NATURAL PONDS AND LAKES are part of nature's water quality treatment, flood protection and stormwater management system. Natural Ponds and Lakes were established in low lying areas and remain with a permanent pool of water at all times. As part of re-development efforts, Natural Pond and Lakes can be incorporated into the stormwater systems to provide treatment, and storm storage and control with certain restrictions. Natural Ponds and Lakes are protected by local, state and federal regulations and, therefore use of these water bodies is subject to strict limitations.

Stormwater treatment efficiency in a Pond or Lake System is related to the confinement time within the Pond or Lake. The longer that stormwater is held in the Pond or Lake, the more nutrients that will be absorbed by biological and chemical processes.

Initial and long-term maintenance costs of Natural Ponds and Lakes consist of removal of exotic (non-native) invasive vegetation in the Pond or Lake and in the surrounding upland preserve.

Natural Pond/lake			
Cost Factor			
	L	M	H
Construction Cost	$		
Maintenance Cost		$	
Stormwater Value			
	L	M	H
Water Quality			X
Storm Control			X
Water Table and Soil Types			
	L	M	H
Water Table	X	X	X
Soil Permeability	X	X	X
L - Low			
M - Medium			
H - High			

Natural Wetlands

Natural Wetlands			
Cost Factor			
	L	M	H
Construction Cost	$		
Maintenance Cost		$	
Stormwater Value			
	L	M	H
Water Quality			X
Storm Control		X	
Water Table and Soil Types			
	L	M	H
Water Table	X	X	X
Soil Permeability	X	X	X
L - Low			
M - Medium			
H - High			

NATURAL WETLANDS have developed naturally over time in locations subject to flooding that as a result have the soil types and water levels that support and maintain wetland vegetation. Wetland areas are a part of nature's stormwater treatment, storage and control system. Existing wetlands are protected and it is important to note that potential impacts to wetland areas are strictly governed by local, state and federal regulations.

Wetlands offer both stormwater storage and treatment. The plant materials absorb water and remove nutrients. They then hold stormwater for an extended time. With careful consideration, existing natural wetland areas can be incorporated within the framework of a stormwater "treatment train" as flow through areas. By incorporating wetland areas it may be possible to recharge and restore an area's natural hydrology. Man-made stormwater treatment areas (STAs) can be designed to mimic natural wetland systems and may also be utilized as part of the stormwater treatment process.

To maintain the health of a wetland, exotic plant materials must be removed. The initial cost to remove exotic plant material can be somewhat expensive depending on the level of infestation. However, on-going maintenance costs of wetlands are minimal, once controlled, and may include annual inspections and removal of any exotic vegetation as part of state or local permitting requirements.

Constructed Wetlands

Constructed Wetlands			
Cost Factor			
	L	M	H
Construction Cost			$
Maintenance Cost		$	
Stormwater Value			
	L	M	H
Water Quality			X
Storm Control			X
Water Table and Soil Types			
	L	M	H
Water Table		X	X
Soil Permeability	X	X	
L - Low			
M - Medium			
H - High			

CONSTRUCTED WETLAND AREAS, also known as stormwater treatment areas (STAs) can provide stormwater storage and treatment by the biological absorption of nutrients by the wetland vegetation. These benefits are similar to Natural Wetland Systems but are not subject to the strict preservation and protection criteria.

To assure viability, these areas are constructed with specific soil types and aquatic vegetation. Constructed Wetland Areas must be carefully designed to consider water table and rainfall patterns.

The construction costs for man-made wetland systems can be high due to the soil amendments and the specific wetland plant species selected. In addition, a temporary irrigation source is sometimes needed to ensure the initial establishment of the plant material. Maintenance costs for the initial grow-in time period can be moderate to high. Once the vegetation is fully established, the on-going maintenance costs for Constructed Wetlands are relatively low and consist of annual or semi-annual inspections of the wetland area and removal of any exotic (non-native) vegetation.

Bio-Retention Swales

Bio-Retention Swales			
Cost Factor			
	L	M	H
Construction Cost		$	
Maintenance Cost	$		
Stormwater Value			
	L	M	H
Water Quality		X	
Storm Control		X	
Water Table and Soil Types			
	L	M	H
Water Table	X	X	
Soil Permeability		X	X
L - Low			
M - Medium			
H - High			

BIO-RETENTION SWALES are similar to standard swales but also include a swale block that acts as a dam to slow the flow of stormwater to retain it for a period of time. The swale blocks in Bio-Retention Swales are typically constructed of earth with sod but can be further stabilized with the use of soil erosion fabric materials or mats.

Although some infiltration and treatment occurs in a standard swale section, the addition of swale blocks in a Bio-Retention Swale system provides for increased storage and infiltration into the underlying soils which results in an overall reduction of runoff and stormwater pollutants. The water quality treatment in a Bio-Retention Swale can be enhanced with the use of certain plant material similar to those used in a Rain Garden.

Bio-Retention Swales are somewhat expensive to construct when compared with typical swales because of the cost of the swale block. On-going maintenance costs are low and consists of mowing and maintenance of plants.

Rain Gardens

RAIN GARDENS are low areas that store and clean stormwater. They function similarly to Dry Retention areas but provide the added benefit of a beautiful garden. Typically the plant choices are ornamental. Rain Gardens can be located in small areas along roadside swale areas, within medians and even in parking lots.

Rain Gardens provide an effective water quality treatment method as well as stormwater storage. Rain Gardens utilize specific plan materials that filter the water along with the soil. Depending on the existing soil types, it may be necessary to add more permeable soils for better infiltration.

Initial costs of Rain Gardens are more expensive than Dry Retention areas because of the select plant materials; however on-going maintenance costs are relatively low. Rain Gardens require periodic maintenance that includes removal of dead and dying plant material and replacing it as needed. Depending on the plant materials selected, a supplemental irrigation source may be required to sustain the Rain Garden areas through the dry season.

Rain Gardens			
Cost Factor			
	L	M	H
Construction Cost		$	
Maintenance Cost	$		
Stormwater Value			
	L	M	H
Water Quality		X	
Storm Control		X	
Water Table and Soil Types			
	L	M	H
Water Table	X	X	
Soil Permeability		X	X
L - Low			
M - Medium			
H - High			

Littoral Zone Planting

Littoral Zone Planting			
Cost Factor			
	L	M	H
Construction Cost		$	
Maintenance Cost		$	
Stormwater Value			
	L	M	H
Water Quality		X	
Storm Control	X		
Water Table and Soil Types			
	L	M	H
Water Table	X	X	X
Soil Permeability	X	X	X
L - Low			
M - Medium			
H - High			

The Littoral Zone of a pond or lake is the portion along the side bank from just above to just below the normal water level. The Littoral area usually consists of an extended shallow shelf planted with native aquatic vegetation both above and below the water level. The planting plan for the Littoral Zone should consider the normal fluctuation of the water level both on a storm by storm basis as well as wet to dry season variations.

Littoral Zone Planting not only creates a functional, efficient way of increasing the water quality treatment but also provides an esthetic enhancement to pond and lake banks. Planting the Littoral Zone can provide additional water quality treatment beyond the level of treatment provided by the lake itself. The biological uptake by the aquatic plant material removes additional nutrients. Littoral Zone Planting also provides additional native habitat both above and below the water line.

The initial costs for Littoral Zone Planting can be relatively high depending upon the plant material selected. On-going maintenance is required to ensure the survivability of the plant materials and to remove and replace dead and dying vegetation to prevent the decomposing plants from releasing the nutrients absorbed by the vegetation back into the pond.

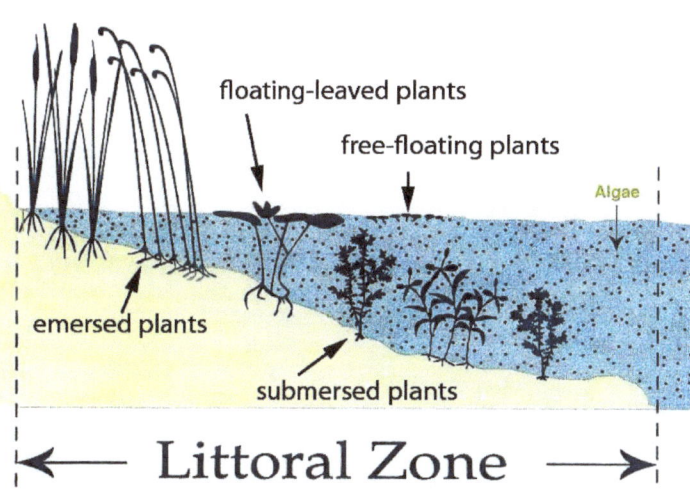

Tree Box Filters

Tree Box Filters			
Cost Factor			
	L	M	H
Construction Cost		$	
Maintenance Cost		$	
Stormwater Value			
	L	M	H
Water Quality	X		
Storm Control	X		
Water Table and Soil Types			
	L	M	H
Water Table	X	X	X
Soil Permeability	X	X	X
L - Low			
M - Medium			
H - High			

TREE BOX FILTERS are similar to manufactured sand filter units but add a vegetation component for both esthetics as well as additional biological treatment. Tree Box Filters typically are individual units designed with specific soil material along with carefully selected trees, shrubs or other types of vegetation. Stormwater is directed into the Tree Box system where it filters down through the soil removing trash, debris, leaves and other matter while the plant material absorbs nutrients. Underdrains collect the treated stormwater after it passes through the filter media and either discharges the treated water to a stormwater pipe or swale system, or is collected in a cistern or other storage system for irrigation or other types of re-use. Due to the relatively small area needed, Tree Box Filters can be installed or retrofitted in urban settings. Tree Box Filters are designed, like other filter systems, for a certain flow rate and usually must have an overflow or by-pass system for larger storm events.

Because of the vegetation aspects of Tree Box Filters, this type of treatment system is more effective than a simple sand filter system in removing dissolved nutrients. Some Tree Box Filters have an initial sedimentation basin area that provides some limited stormwater storage but typically, storm storage and control must be provided in other facilities.

Costs for installation are low to moderate for new construction but may be higher for retrofit areas where existing facilities are affected. As with other types of filters and stormwater inlet devices, periodic maintenance is required to remove accumulated debris. Maintenance of the vegetation is also required.

Algae Mat

Algae Mat Treatment			
Cost Factor			
	L	M	H
Construction Cost			$
Maintenance Cost			$
Stormwater Value			
	L	M	H
Water Quality			X
Storm Control	X		
Water Table and Soil Types			
	L	M	H
Water Table	X	X	X
Soil Permeability	X	X	X
L - Low			
M - Medium			
H - High			

ALGAE MATS are a method of water quality treatment that uses biological absorption of nutrients by various forms of algae. The Algae Mat system is based on pumping a controlled thin layer of stormwater over a relatively flat area covered with a mat where algae have been established. The algae grow vigorously under these conditions, absorbing the nutrients from the stormwater.

Since the Algae Mats themselves do not provide storage of stormwater, this method of treatment must be coupled with storage and pumping facilities. Depending of the flow volume to be treated, a relatively large area is needed for the algae mats to allow for the slow sheets of stormwater flow necessary to effectively remove nutrients.

The maintenance costs are high as the algae must be harvested and removed as often as once a week. The harvested algae are not wasted. It can be composted into a soil enhancement material or cattle feed. The harvested algae can also be marketed as a source for conversion into bio-fuels.

The initial costs for an Algae Mat Treatment System are high. The initial and long-term costs make the use of Algae Mat Treatment systems unpractical for smaller applications.

Green Roof

Green Roofs			
Cost Factor			
	L	M	H
Construction Cost			$
Maintenance Cost		$	
Stormwater Value			
	L	M	H
Water Quality			X
Storm Control		X	
Water Table and Soil Types			
	L	M	H
Water Table	X	X	X
Soil Permeability	X	X	X
L - Low			
M - Medium			
H - High			

Credit: Kimley Horn Associates

Media and Vegetation
Drainage Layer and Seperation Fabric
Insulation
Protection Layer
Water Proofing
Primer
Roof Surface

GREEN ROOFS feature a roof-top garden that provides stormwater treatment and storage. Carefully selected vegetation and soils collect and clean rainwater by absorbing nutrients. It is recommended that Green Roofs consist of Florida native vegetation and drought tolerant plants. When planning for a Green Roof, the design should be made to accommodate the additional weight from the soil and plant materials. Also, the Green Roof must be designed to prevent wind uplift of the soil media and vegetation. In order to insure the long-term viability of the Green Roof vegetation, an irrigation system should be included in the Green Roof design.

The initial construction cost of a Green Roof is greater than a conventional roof because of the additional structural requirements and cost to install the soil, plant and irrigation. Studies have indicated that additional cost for Green Roof systems can be recovered over time due through savings of heating and air conditioning costs as these roofs offer an insulating affect of the soil and plant materials. Green Roofs also have a longer lifespan than conventional roof systems. Maintenance requirements for Green Roofs are similar to other landscape areas and include periodic inspections of the plant material, irrigation system and the roof drainage system. Dead and dying plants should be removed along with weeds and other non-native plants. It also may be necessary to supplement the soil periodically.

Credit: Kimley Horn Associates

Vegetated Filter Strips

VEGETATED FILTER STRIPS are long, linear landscaped areas along roadsides. Vegetated Filter Strips are similar to Rain Gardens but they lack the aesthetic plant material found in Rain Gardens. Vegetated Filter Strips are often an initial part of a "Treatment Train" and are combined with other stormwater practices to provide enhanced treatment and storage of stormwater.

Due to the limited width, Filter Strip areas are usually shallow and provide limited stormwater storage. Rather than discharging stormwater from paved and roofed surfaces directly into inlets, runoff can be diverted into and across the Filter Strip providing filtration and treatment by the plant materials.

Like other landscaped areas, Filter Strips require periodic maintenance. A supplemental irrigation source may be needed during the dry season, especially when drought-tolerant native plant species are not utilized.

Vegetated Filter Strips			
Cost Factor			
	L	M	H
Construction Cost	$		
Maintenance Cost	$		
Stormwater Value			
	L	M	H
Water Quality	X		
Storm Control	X		
Water Table and Soil Types			
	L	M	H
Water Table	X	X	
Soil Permeability		X	X
L - Low			
M - Medium			
H - High			

Exfiltration Trench

EXFILTRATION TRENCHES consist of stormwater facilities that direct runoff to an underground system that allows water to soak into the surrounding soils. Exfiltration Trench systems include surface inlets connected to an underground piping system that is surrounded by rock material. Exfiltration system pipes may be concrete, metal or plastic with slots or perforations that allow water to flow out of the pipe into the rock and then into the soil.

Stormwater storage is provided within the pipe itself and also in the void space within the rock material. Exfiltration Trenches function best in areas with highly permeable soils and with a deep depth to the water table. Exfiltration Trenches provide an effective water quality treatment method by holding stormwater runoff and allowing the soils to filter it, thereby reducing discharges to surface waters and reducing pollutant loadings.

Construction of Exfiltration Trenches is generally more expensive than surface storage. But this expense can be offset by providing more developable land. Maintenance is moderate and includes periodic observations to assure the Exfiltration Trenches are functioning properly. Vacuuming is also required to remove sediments, debris and trash that accumulate.

Exfiltration Trenches			
Cost Factor			
	L	M	H
Construction Cost		$	
Maintenance Cost		$	
Stormwater Value			
	L	M	H
Water Quality			X
Storm Control		X	
Water Table and Soil Types			
	L	M	H
Water Table	X	X	
Soil Permeability		X	X
L - Low			
M - Medium			
H - High			

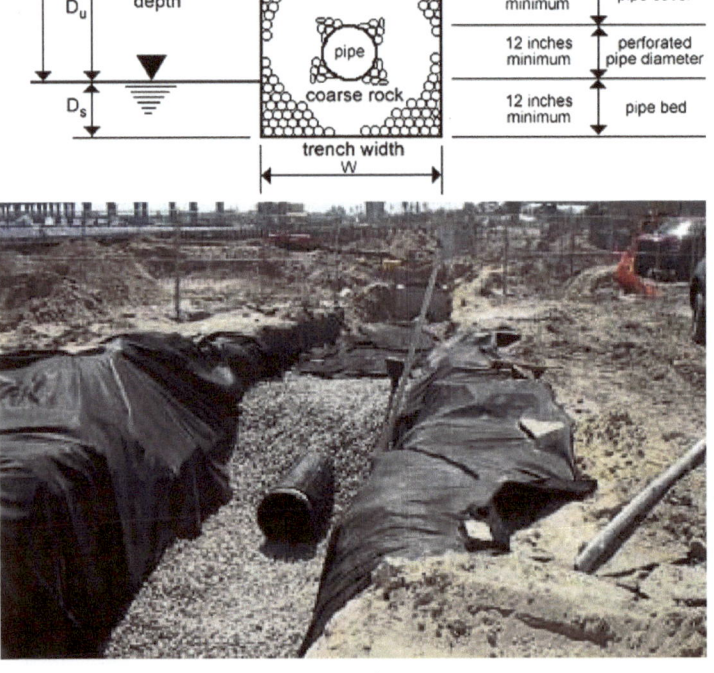

TYPICAL EXFILTRATION TRENCH

Sand Filters

Sand Filters			
Cost Factor			
	L	M	H
Construction Cost		$	
Maintenance Cost		$	
Stormwater Value			
	L	M	H
Water Quality		X	
Storm Control	X		
Water Table and Soil Types			
	L	M	H
Water Table	X	X	X
Soil Permeability	X	X	X
L - Low			
M - Medium			
H - High			

SAND FILTER systems have been used for treatment of potable water, wastewater and stormwater for many years. Typically, Sand Filters for stormwater systems operate by gravity and include specifically graded clean sands placed in a basin with an inflow control and an underdrain system. Stormwater is directed onto the surface and is filtered as it flows through the sand media. Sand Filters can be located on the surface or can be constructed in underground chambers. Usually some type of upstream pre-treatment screen, baffle and/or sedimentation chamber is included in the Filter design to remove trash, leaves, debris and larger sediments that would quickly clog the pores in the sand media. Sand Filters are designed for a specific maximum flow rate and therefore bypass and overflow facilities must be included for larger or more intense storm events.

Sand Filters can be constructed in many different sizes and shapes to accommodate specific site requirements. There are several manufacturers of pre-fabricated Sand Filter units with or without sedimentation basins or as an alternative, Sand Filters can be constructed in place.

Construction costs for Sand Filters can be relatively low to moderately expensive depending on the size of the system and type of structure used and whether the system is constructed on the surface or underground. Maintenance cost is moderate and includes removal and replacement of the filter media

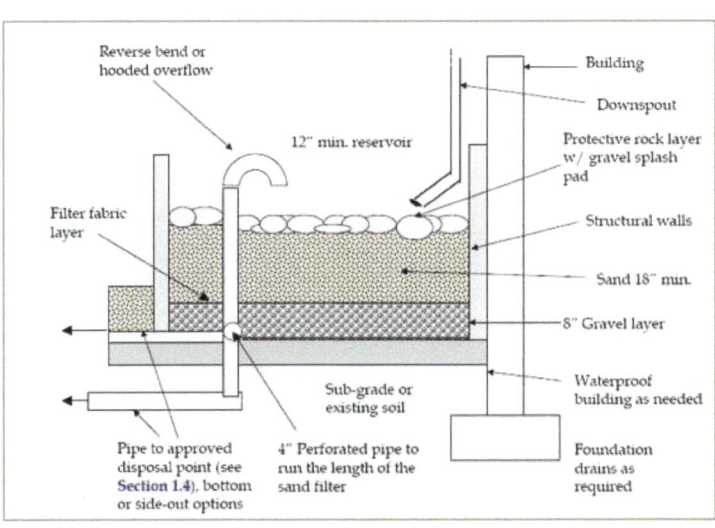

Rock Soaking Area

ROCK SOAKAGE AREAS are often used to enhance the storage and infiltration capabilities of existing soils. These Areas are constructed at the ground surface but can be at the bottom of a depressed area. As stormwater flows across the rock surface, it fills the void spaces between the rocks and is filtered by the surrounding soils.

Rock Soakage Areas are an effective way to reduce stormwater pollutants by intercepting stormwater before it goes off-site and filtering it.

Rock Soakage Areas are reasonably inexpensive to construct, and maintenance is minimal. However, as the void spaces in the rock become clogged, it may be necessary to remove some or all of the rock and either replace it with new material or wash out the fine silt and sand before returning it to the Soakage Area.

Rock Soakage Area			
Cost Factor			
	L	M	H
Construction Cost	$		
Maintenance Cost	$		
Stormwater Value			
	L	M	H
Water Quality		X	
Storm Control		X	
Water Table and Soil Types			
	L	M	H
Water Table	X	X	
Soil Permeability		X	X
L - Low			
M - Medium			
H - High			

Floating Vegetated Mats

FLOATING VEGETATED MATS are an efficient way of increasing the water quality treatment capacity within a wet retention/detention system. Floating Vegetated Mats are a system of mat material that floats on the water surface of a pond or lake. The Mats are anchored to the lake bottom or along the side of the lake and can rise and fall with the water level. Aquatic plants are carefully selected and inserted in openings in the Mats with the root systems extended into the water. As these plants grow, the submerged root systems absorb and remove nutrients biologically.

The Mats not only improve the water quality in a lake system but also create additional habitat for birds and fish.

The initial and annual replacement costs are relatively high for the Floating Mat and plant materials depending on the size of the system. Once installed, there is little on-going maintenance needed. The plant material is harvested and replaced with new plant material once or twice a year. The harvested plant material is not wasted. It can be composted and used to enrich soil.

Floating Vegetated Mats			
Cost Factor			
	L	M	H
Construction Cost			$
Maintenance Cost			$
Stormwater Value			
	L	M	H
Water Quality			X
Storm Control	X		
Water Table and Soil Types			
	L	M	H
Water Table	X	X	X
Soil Permeability	X	X	X
L - Low			
M - Medium			
H - High			

Vegetated Wall

VEGETATED or **GREEN WALL SYSTEMS** are similar to Green Roof Systems but are constructed vertically along a wall surface. Supporting framework is needed to hold the soil and vegetation in place. The vegetated wall is watered through stormwater that flows down through the soil/plant matrix from roof areas.

Vegetated Walls offer both stormwater storage and treatment. The soil reduces overall runoff volumes because it absorbs and filters the rainwater as it trickles down the wall system. Plant materials must be carefully selected to assure that the plants selected can thrive on the wall surface and that the plants provide maximum absorption of nutrients and other pollutants.

The construction cost for a vegetated wall is more expensive than a conventional wall because of the added costs for structural framing, soil, and vegetation and irrigation system. However, the vegetated wall provides insulating benefits which offset the additional cost. Cost for on-going maintenance is relatively moderate but becomes more difficult as the wall height increases.

Maintenance includes regular inspections of the landscape area and routine maintenance of plant material, soil and irrigations system.

Vegetated Wall			
Cost Factor			
	L	M	H
Construction Cost			$
Maintenance Cost		$	
Stormwater Value			
	L	M	H
Water Quality		X	
Storm Control	X		
Water Table and Soil Types			
	L	M	H
Water Table	X	X	X
Soil Permeability	X	X	X
L - Low			
M - Medium			
H - High			

Stormwater Filters

Stormwater Filters			
Cost Factor			
	L	M	H
Construction Cost		$	
Maintenance Cost		$	
Stormwater Value			
	L	M	H
Water Quality		X	
Storm Control	X		
Water Table and Soil Types			
	L	M	H
Water Table	X	X	X
Soil Permeability	X	X	X
L- Low			
M - Medium			
H - High			

There are many different types of **STORMWATER FILTER** units. Some Filters are designed for specific types of stormwater pollutants (i.e. oils, grease and other hydrocarbons) while other types of Filter units are for general removal of sands, silts and other matter. There are different types of filter media used including fabrics, sand media as well as special proprietary media for specific stormwater constituents.

Stormwater Filters are quite effective in removing sediments, trash and debris from stormwater runoff but must be maintained on a regular basis to prevent the accumulated materials from simply passing through when the capacity is reached. Filter systems are designed with an overflow or by-pass to allow runoff to pass during larger storm events. The by-pass also allows flows to continue if the Filters become clogged. Although Stormwater Inlet Devices can help keep significant amounts of sediments and organic matter out of stormwater discharges, these types of systems generally have a very small effect on dissolved nutrients.

The cost for installation of Stormwater Filter systems is moderately expensive depending on the type of filter and whether it is a complete unit or an insert into an existing structure. Stormwater Filters require regular inspections and removal of accumulated sand, sediment and debris. Maintenance cost is moderate and is done by hand or with a vacuum system. It can be necessary to periodically replace the Filter units or if the Filter utilizes sand or other media, remove and replace the Filter media.

Inlet Screens, Baffles, Sumps

Screens, Baffles, Sumps			
Cost Factor			
	L	M	H
Construction Cost	$		
Maintenance Cost		$	
Stormwater Value			
	L	M	H
Water Quality	X		
Storm Control	X		
Water Table and Soil Types			
	L	M	H
Water Table	X	X	X
Soil Permeability	X	X	X
L - Low			
M - Medium			
H - High			

Standard stormwater inlets are concrete structures with a surface grate and/or slot at the surface designed to collect stormwater and release it through a pipe system. By adding SCREENS, BAFFLES, AND SUMPS to the inlet, debris and sediments can be captured and kept out of the stormwater stream. Screens for inlets are usually made of metal or plastic with mesh openings that allow water to pass but collect larger sized debris. A baffle is a vertical covering across the pipe opening that extends below the bottom of the pipe to keep floating debris out of the pipe system. A sump is an extended basin at the bottom of the inlet where some of the sands and larger sediments can settle out and collect.

Screens, Baffles and Sumps are the initial part of a stormwater treatment train and improve water quality by removing larger particles upstream of other stormwater facilities. The treatment level of Screens, Baffles and Sumps is limited and suspended solids, nutrients and other stormwater pollutants generally pass through these inlet devices.

Inlet Screens, Baffles and Sumps are a relatively inexpensive to construct and can be easily retrofitted into existing structures. Maintenance costs are moderate and consist of regular inspections and removal of accumulated debris and sediments by hand in shallower inlets, or by a vacuum truck for deeper or less accessible inlet structures.

Baffle Boxes

Baffle Boxes			
Cost Factor			
	L	M	H
Construction Cost		$	
Maintenance Cost			$
Stormwater Value			
	L	M	H
Water Quality	X		
Storm Control	X		
Water Table and Soil Types			
	L	M	H
Water Table	X	X	X
Soil Permeability	X	X	X
L - Low			
M - Medium			
H - High			

BAFFLE BOXES are pre-cast (or cast-in-place) underground tanks usually made of concrete that are separated with internal valve walls into two or more compartments. The inlet piping in a typical Baffle Box is normally fitted with a screen device to separate out large debris, leaves and trash. The Baffles trap silts, sands and sediments and allow them to settle to the bottom of the box. The internal valves also prevent floating materials that may pass through the screen from discharging out of the system. Baffle Box systems include a by-pass system also allows stormwater flows to continue through the system if screens and baffles become clogged with debris.

Baffle Boxes are often retrofitted to the downstream section of an existing stormwater treatment train as a water quality treatment method. These types of systems are very effective in removing silt and sediment as well as floating debris from stormwater runoff but are not a very efficient method for removing dissolved pollutants and nutrients.

Construction costs are moderate, and there are several manufacturers of pre-fabricated Baffle Box systems. Maintenance costs can be high and include monthly inspections and vacuuming as needed to remove trash and debris from the screens. Typically this is done with a vacuum truck system.

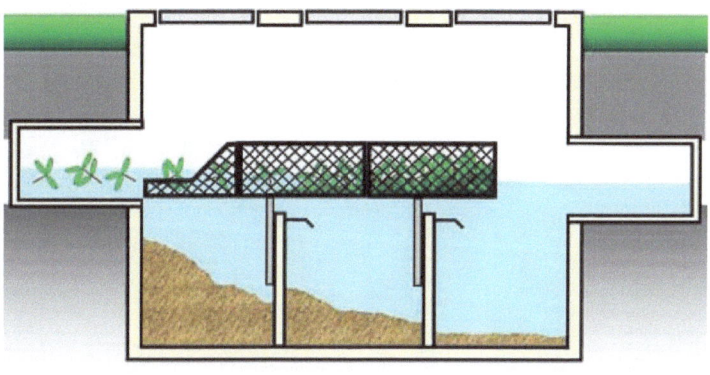

Vortex Separators

VORTEX SEPARATORS are a type of manufactured stormwater device for the removal of sand and sediments from stormwater runoff. The circular design of vortex systems creates a swirling pattern that forces sediments and debris to the outside perimeter by centrifugal force. The separated materials then settle into an attached chamber where they are held for eventual removal. Typically, Vortex Separators have an overflow or by-pass system to allow runoff to pass during larger storm events.

Vortex Separators are effective in removing sand, sediments, trash, debris and larger organic matter (leaves, etc.) from stormwater runoff but have little removal effects on dissolved solids and nutrients.

Vortex Separators are moderately expensive to install and require regular inspections and removal of accumulated sand, sediment and debris. Installation can be as attachments to existing systems or as separate units in new construction. Typically the maintenance of these devices is done with a vacuum system.

Vortex Separators			
Cost Factor			
	L	M	H
Construction Cost		$	
Maintenance Cost		$	
Stormwater Value			
	L	M	H
Water Quality	X		
Storm Control	X		
Water Table and Soil Types			
	L	M	H
Water Table	X	X	X
Soil Permeability	X	X	X
L - Low			
M - Medium			
H - High			

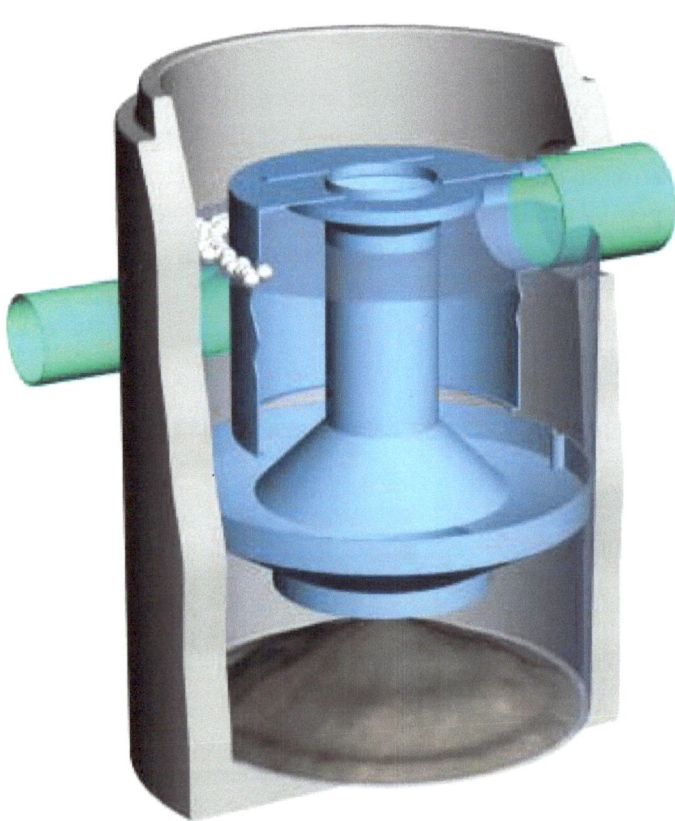

Chemical (Alum) Treatment

Various **CHEMICAL TREATMENT SYSTEMS** have been used over the years for treatment of stormwater and drinking water. The injection of alum (aluminum sulfate) into stormwater runoff is the most common chemical used for stormwater treatment. Once mixed into the stormwater, these chemicals combine with nutrients, suspended solids, heavy metals and other materials to create larger particles that can settle out of the water column. The effectiveness is related to the confinement time in the system. This type of treatment can be done on a small scale with settling tanks but for stormwater, the chemical treatment is often injected prior to larger stormwater retention/detention systems.

Although chemical treatment of stormwater is an effective method for reducing pollutants in stormwater, this method is not often used on small scale operations.

The initial costs for chemical treatment of stormwater can be expensive, requiring carefully controlled pumps and monitoring systems to insure correct application. There are also relatively high on-going costs for chemical materials as well as operation and maintenance.

Chemical Treatment			
Cost Factor			
	L	M	H
Construction Cost			$
Maintenance Cost			$
Stormwater Value			
	L	M	H
Water Quality			X
Storm Control	X		
Water Table and Soil Types			
	L	M	H
Water Table	X	X	X
Soil Permeability	X	X	X
M - Medium			
H - High			

Modules

STORMWATER DESIGN TOOLKIT
Martin County Community Redevelopment Agency

How to Use

DESIGN MODULES illustrate stormwater management within typical roadway sections found in the Martin County Community Redevelopment Areas. The series of Design Modules consist of three roadway sizes with uses defined as typical, best management and innovative. The expected outcome would be to provide examples of using stormwater tools that inspire future redevelopment. In each case, stormwater treatment and storage is demonstrated above and below ground through visual graphics and supplementary explanation of key development issues that may rise out of each Design Module.

As with all development, land use is optimized to provide higher rates of return by implementing smart stormwater tools. Density may increase creating higher investment opportunities while providing the community a place to flourish.

PAVEMENT
PERVIOUS PAVERS

Applied Strategies: *These keys link the strategies in the sections to the strategies identified in the report*

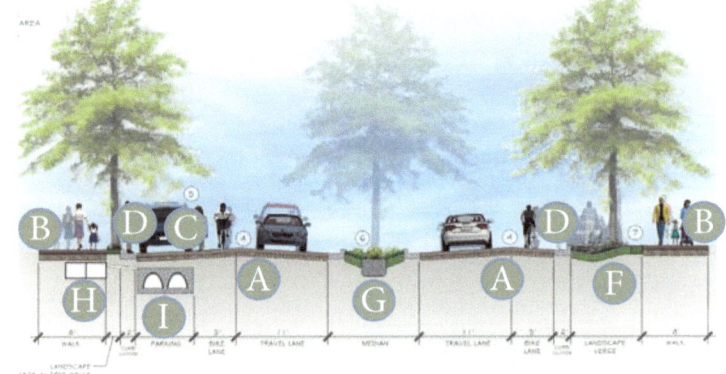

Street Section: *Typical sections to deomonstrate implementation*

Perspectives: *To deomonstrate the various Innovative Management Strategies. These illustrative perspectives provide a clear vision on how each various strategies are applied*

Martin County Board of County Commissioners
Your County, Your Community

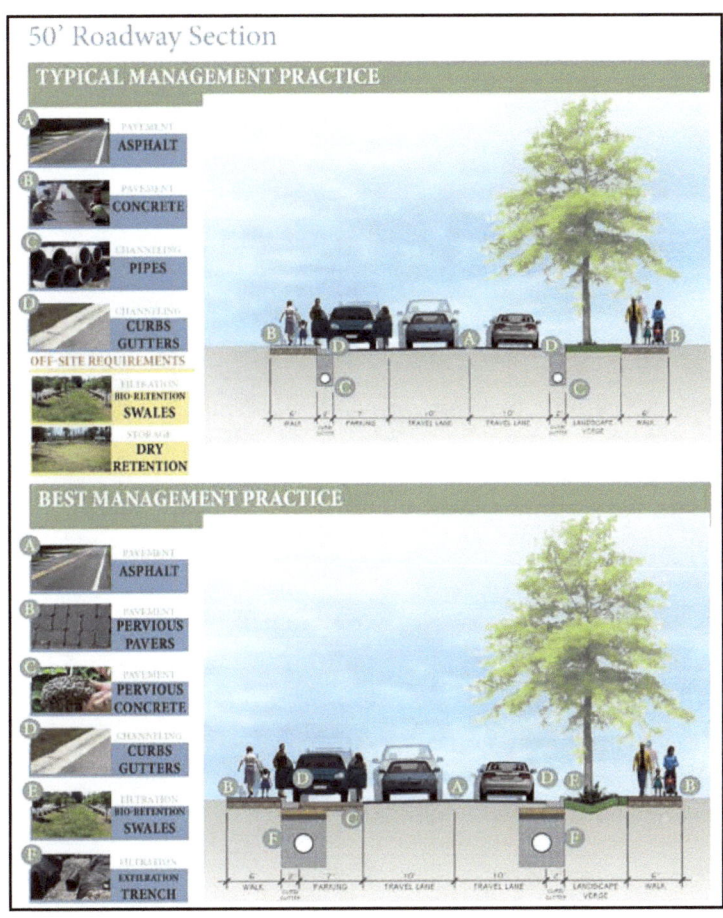

STREET SECTIONS

Street sections are presented in three ways: typical design, best management practice which illustrates use of some sustainable practices, and innovative management which illustrates use of several sustainable tools combined. The report illustrates three typical right of way widths: 50 feet, 80 feet and 100 feet.

50' Roadway Section

TYPICAL MANAGEMENT PRACTICE

- A — PAVEMENT — ASPHALT
- B — PAVEMENT — CONCRETE
- C — CHANNELING — PIPES
- D — CHANNELING — CURBS GUTTERS

OFF-SITE REQUIREMENTS

- FILTRATION — BIO-RETENTION SWALES
- STORAGE — DRY RETENTION

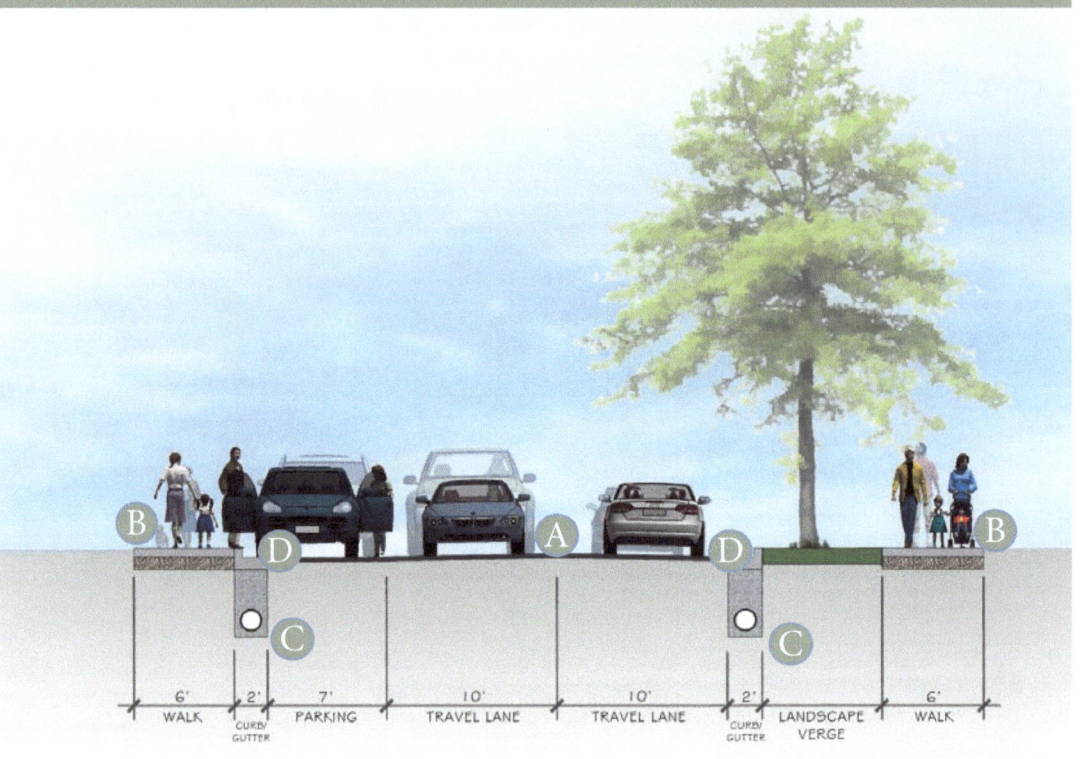

BEST MANAGEMENT PRACTICE

- A — PAVEMENT — ASPHALT
- B — PAVEMENT — PERVIOUS PAVERS
- C — PAVEMENT — PERVIOUS CONCRETE
- D — CHANNELING — CURBS GUTTERS
- E — FILTRATION — BIO-RETENTION SWALES
- F — FILTRATION — EXFILTRATION TRENCH

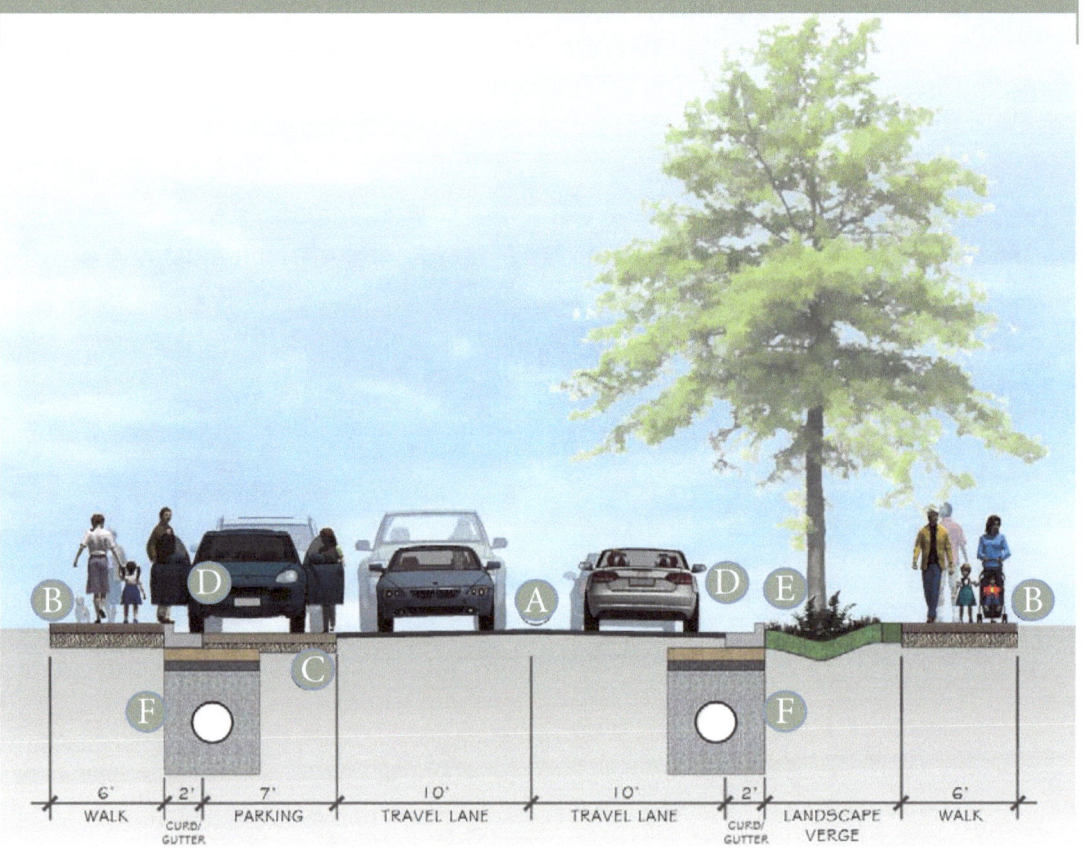

Martin County Board of County Commissioners
Your County, Your Community

50' Roadway Section
Innovative Management

APPLIED STRATEGIES

A — PAVEMENT — **PERVIOUS ASPHALT**

B — PAVEMENT — **PERVIOUS PAVERS**

C — PAVEMENT — **PERVIOUS CONCRETE**

D — CHANNELING — **CURBS GUTTERS**

E — FILTRATION — **BIO-RETENTION SWALES**

F — FILTRATION — **TREE BOX**

G — STORAGE — **PLASTIC ARCH CHAMBERS**

H — FILTRATION — **EXFILTRATION TRENCH**

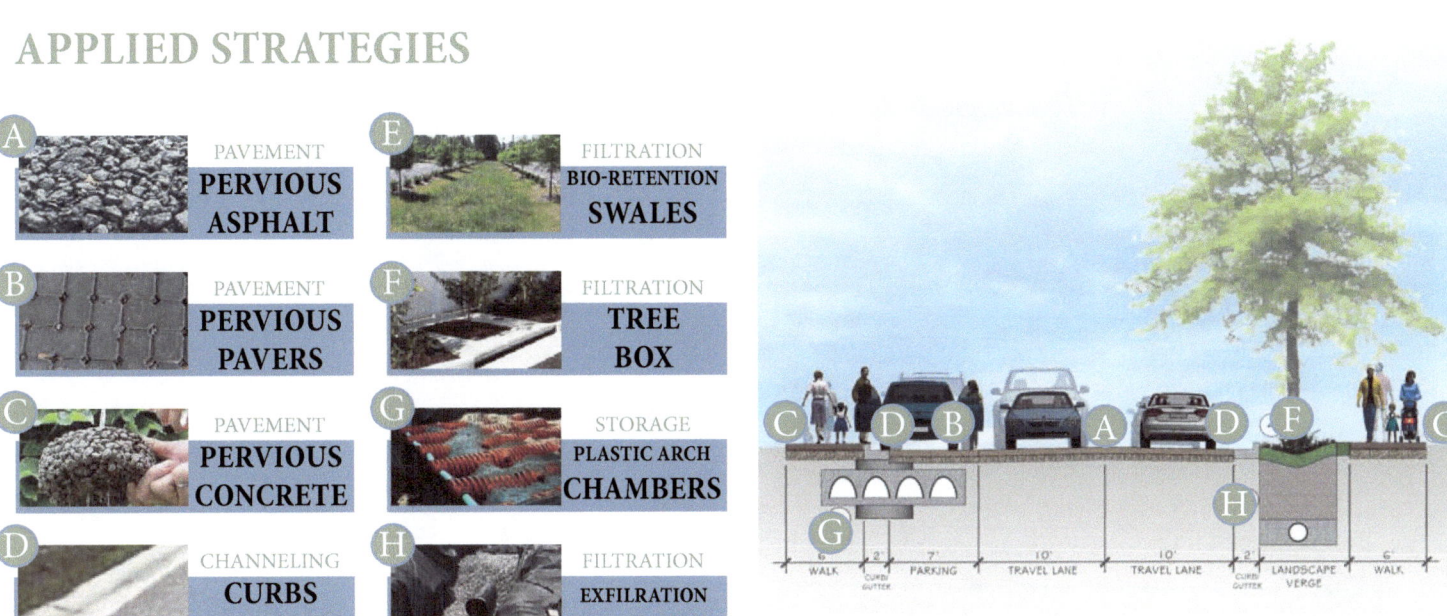

Martin County Community Redevelopment Agency
Redevelopment in Action

80' Roadway Section

TYPICAL MANAGEMENT PRACTICE

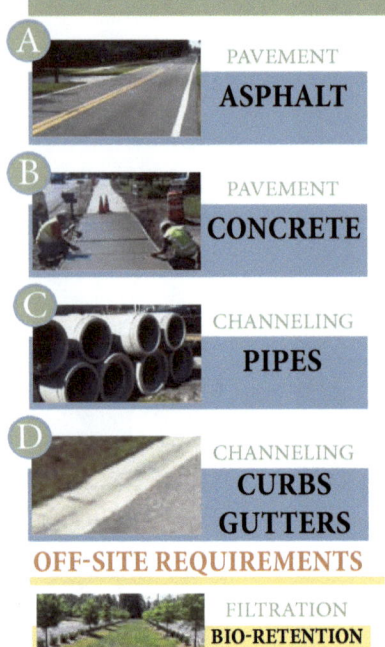

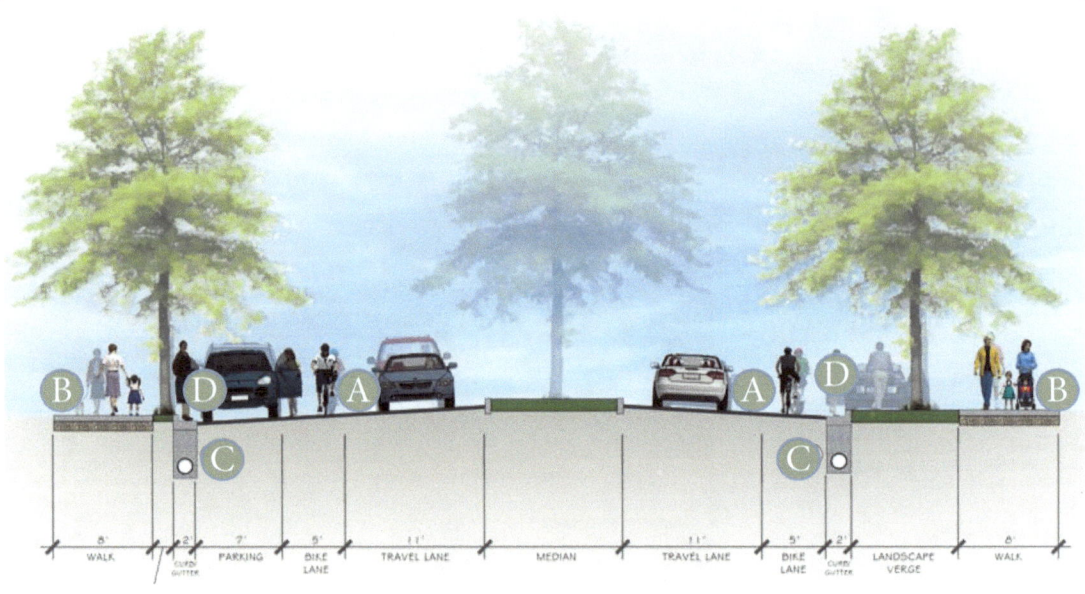

A — PAVEMENT — **ASPHALT**
B — PAVEMENT — **CONCRETE**
C — CHANNELING — **PIPES**
D — CHANNELING — **CURBS GUTTERS**

OFF-SITE REQUIREMENTS

FILTRATION — **BIO-RETENTION SWALES**
STORAGE — **DRY RETENTION**

BEST MANAGEMENT PRACTICE

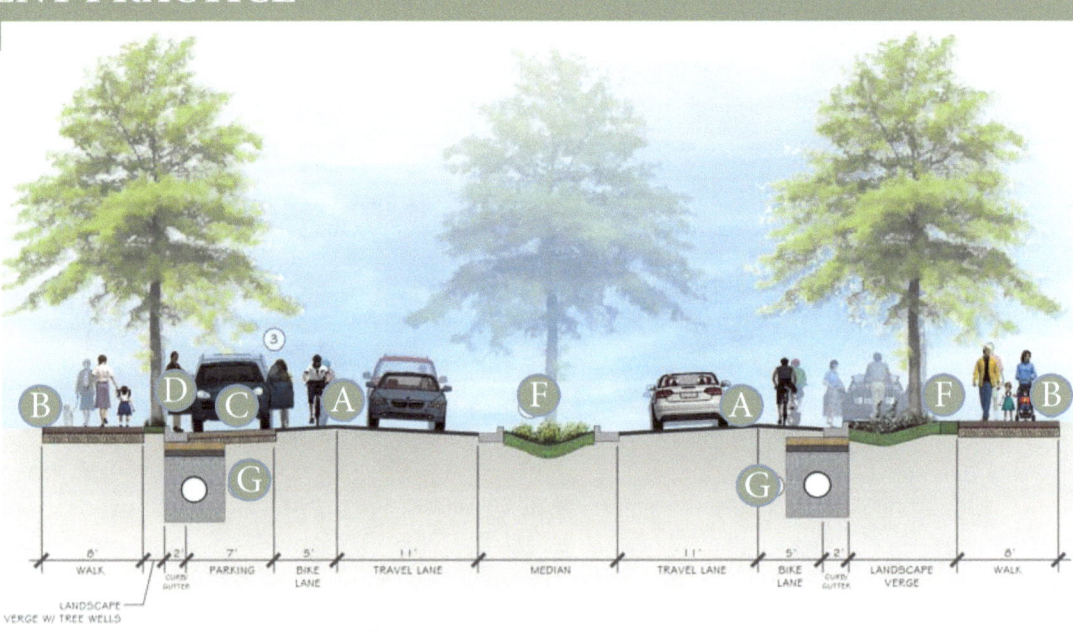

 A — PAVEMENT — **ASPHALT**
 B — PAVEMENT — **PERVIOUS PAVERS**
 C — PAVEMENT — **PERVIOUS CONCRETE**
 D — CHANNELING — **CURBS GUTTERS**
 E — CHANNELING — **PIPES**
 F — FILTRATION — **BIO-RETENTION SWALES**
 G — FILTRATION — **EXFILTRATION TRENCH**

Martin County Board of County Commissioners
Your County, Your Community

80' Roadway Section
Innovative Management

APPLIED STRATEGIES

A PAVEMENT — PERVIOUS ASPHALT

B PAVEMENT — PERVIOUS PAVERS

C PAVEMENT — PERVIOUS CONCRETE

D CHANNELING — CURBS GUTTERS

E CHANNELING — PIPES

F FILTRATION — BIO-RETENTION SWALES

G FILTRATION — TREE BOX

H STORAGE — CONCRETE CHAMBERS

I STORAGE — PLASTIC ARCH CHAMBERS

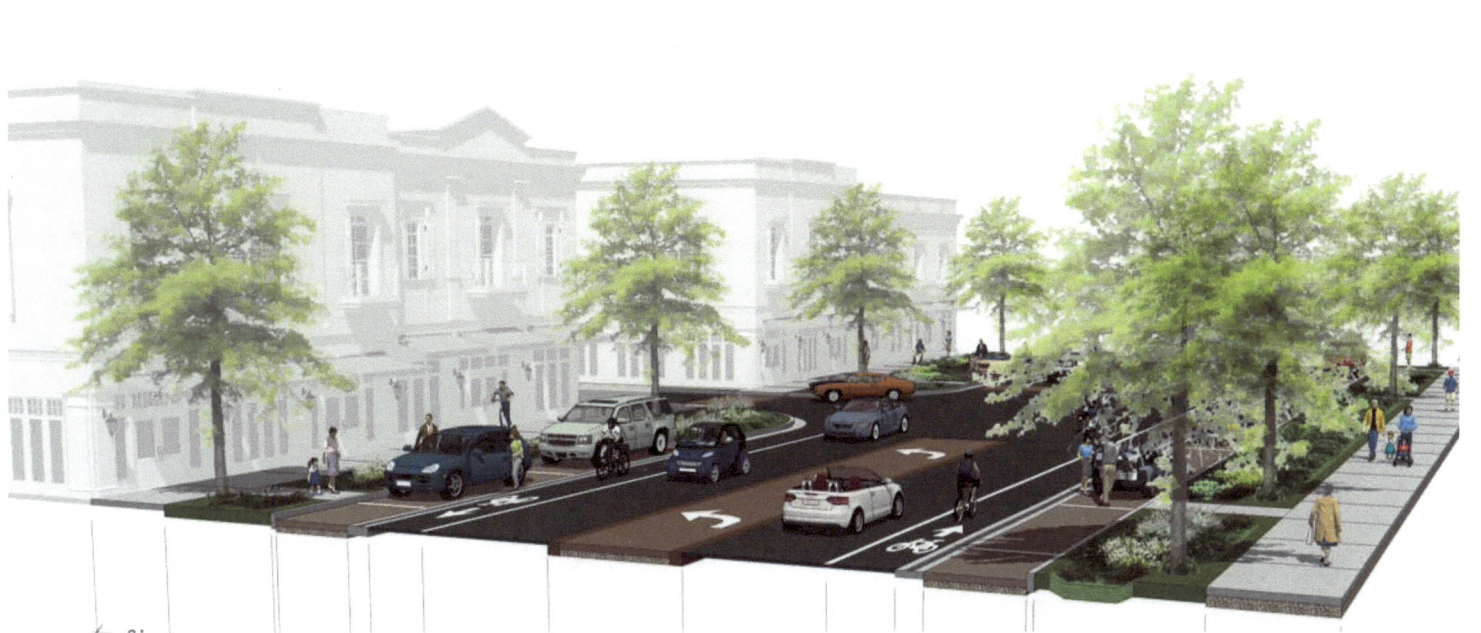

100' Roadway Section

TYPICAL MANAGEMENT PRACTICE

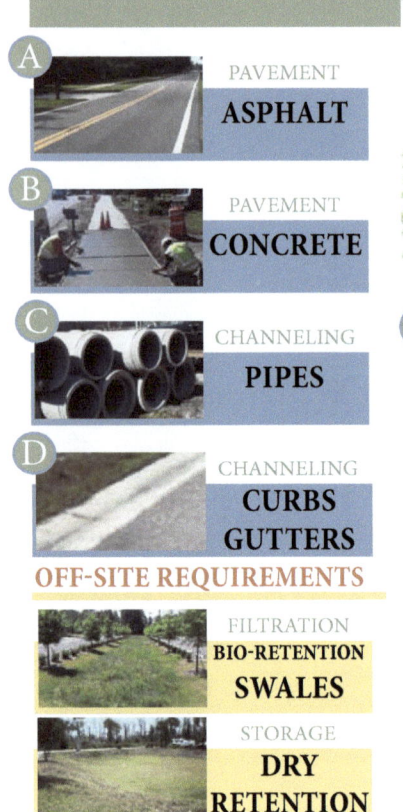
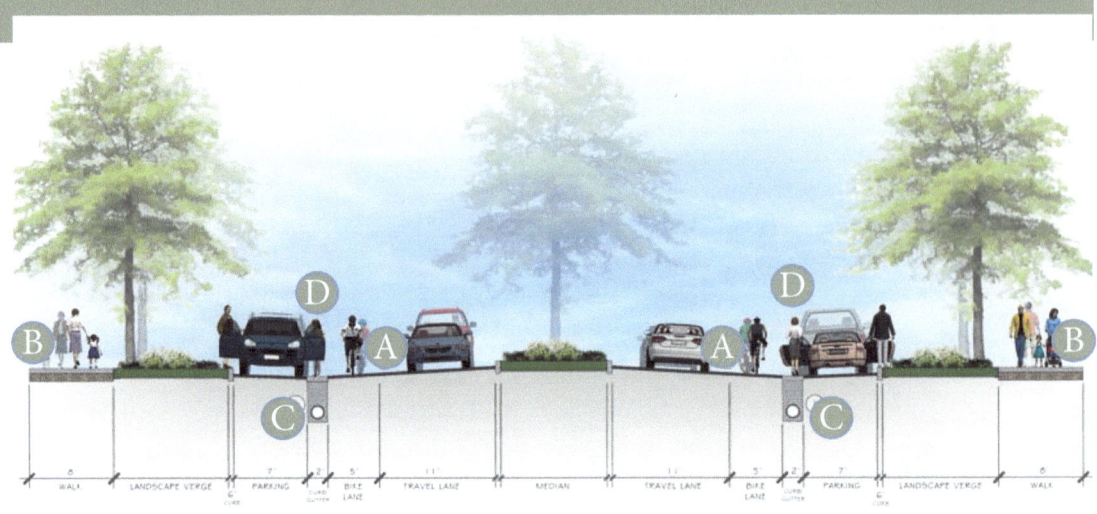

- A — PAVEMENT: **ASPHALT**
- B — PAVEMENT: **CONCRETE**
- C — CHANNELING: **PIPES**
- D — CHANNELING: **CURBS GUTTERS**

OFF-SITE REQUIREMENTS

- FILTRATION: **BIO-RETENTION SWALES**
- STORAGE: **DRY RETENTION**

BEST MANAGEMENT PRACTICE

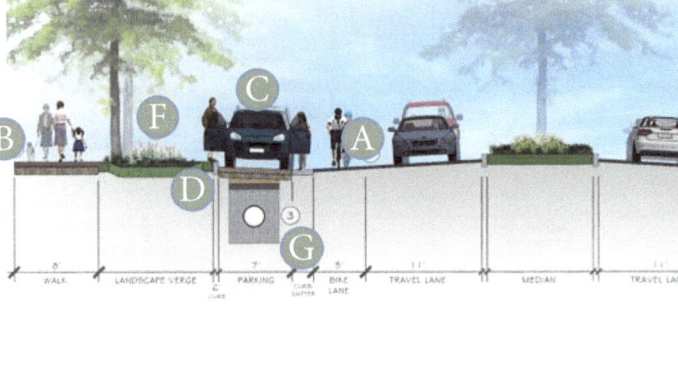

 A — PAVEMENT: **ASPHALT**

 B — PAVEMENT: **PERVIOUS PAVERS**

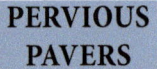

 C — PAVEMENT: **PERVIOUS CONCRETE**

 D — CHANNELING: **CURBS GUTTERS**

 E — CHANNELING: **PIPES**

 F — FILTRATION: **RAIN GARDEN**

 G — FILTRATION: **EXFILTRATION TRENCH**

Martin County Board of County Commissioners
Your County, Your Community

100' Roadway Section
Innovative Management

APPLIED STRATEGIES

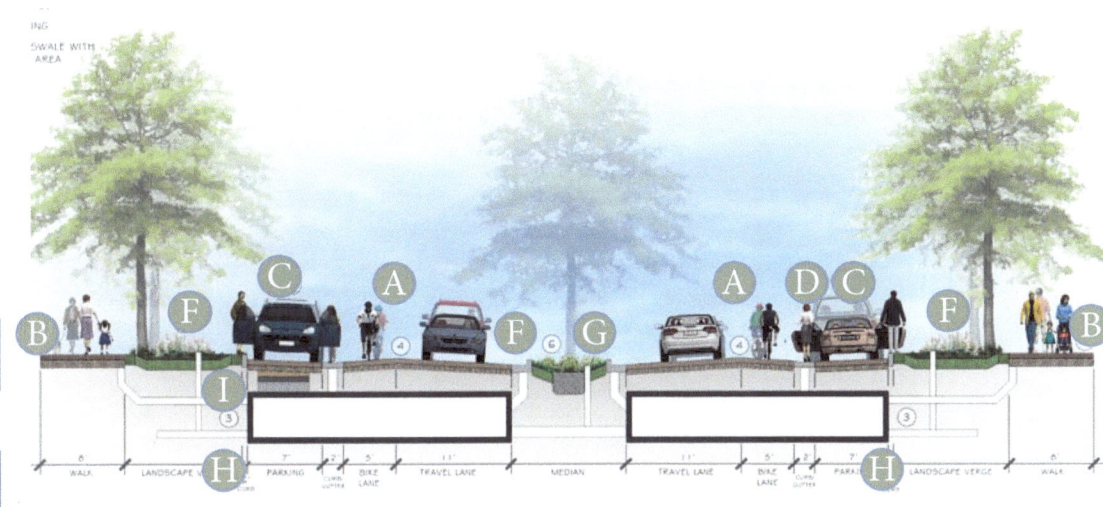

A — PAVEMENT **PERVIOUS ASPHALT**

B — PAVEMENT **PERVIOUS PAVERS**

C — PAVEMENT **PERVIOUS CONCRETE**

 **D** — CHANNELING **CURBS GUTTERS**

 **F** — FILTRATION **RAIN GARDEN**

 **H** — STORAGE **UNDER GROUND**

 **E** — CHANNELING **PIPES**

 **G** — FILTRATION **SOAKING AREA**

 **I** — FILTRATION **EXFILTRATION TRENCH**

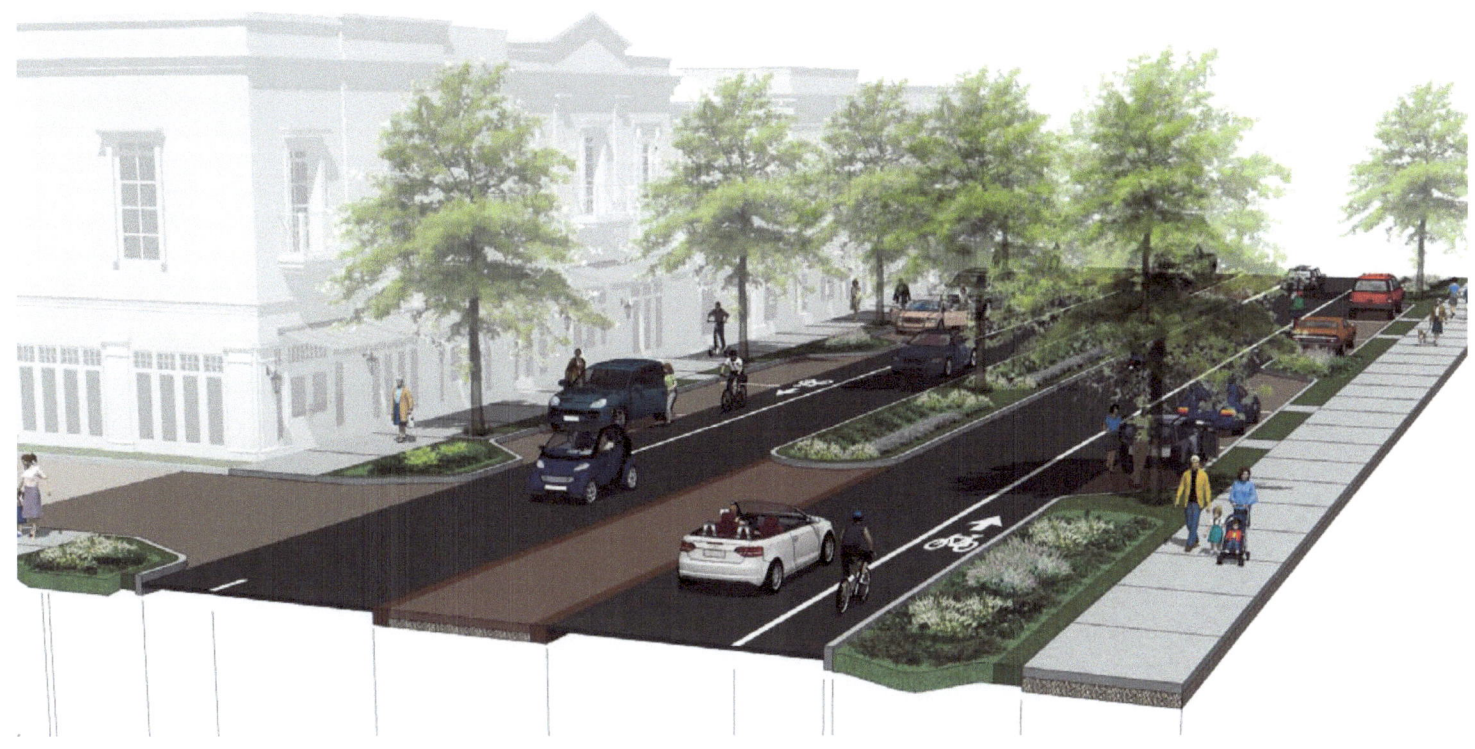

Martin County Community Redevelopment Agency
Redevelopment in Action

Implementation

STORMWATER DESIGN TOOLKIT
Martin County Community Redevelopment Agency

Mixing Materials

The stormwater management tools presented in this report provides a broad range of materials and techniques to manage stormwater. These tools can be combined and mixed to not only achieve maximum stormwater management, but also infinite possibilities and character.

Once the desires for cost, soil types, and treatment options, have been established, then multiple materials and tools can be combined. This allows for variety, and the creation of unique places. This will not only create improved water quality in the Martin County Community Redevelopment Areas, but will also provide the natural variety that makes each one of our Redevelopment Areas unique.

Below are examples of projects that combined materials to maximize the management and treatment of stormwater. Each provides a unique atheistic character.

Martin County Board of County Commissioners
Your County, Your Community

Next Steps

This is a living document, and it will be updated as stormwater management systems evolve. In preparing this document, the staff of the Community Development Department was inspired to continue the process and explore ways to demonstrate lessons that were learned in compiling the report. It is our intent to:

- Review and organize for endorsement from South Florida Water Management District
- Collaborate with Martin County Departments to update standards and regulations to allow implementation by developers
- Present to Martin County Community Redevelopment Agency for adoption
- Present to Martin County Board of County Commissioners for adoption
- Seek alternative funding mechanisms to implement demonstration projects
- Partner with a local developer and manufacturer of these systems for demonstrations and research data as required
- Provide educational sessions to the public on utilizing these strategies
- Implement innovative strategies into a pilot project

Acknowledgements

This report was compiled utilizing expertise of several community partners whom without their resources and support could not have been possible. The Community Development would like to acknowledge the following:

COMMUNITY DEVELOPMENT STAFF
Kev Freeman, Director
Bonnie Landry, AICP, CNU-A, Community Development Specialist
Sarah Rose Henke, PE, LEED, Community Redevelopment Project Engineer
Nakeischea Smith, AICP, Community Development Specialist
Edward Erfurt, Urban Designer

SOUTH FLORIDA WATER MANAGEMENT DISTRICT

MARTIN COUNTY BOARD OF COUNTY COMMISSIONERS

KIMLEY-HORN AND ASSOCIATES, INC.

CREECH ENGINEERS, INC

D P Z **DUANY PLATER-ZYBERK & COMPANY**

UNIVERSITY OF CENTRAL FLORIDA

JULIE PREAST, ED PREAST, INC.

VERTEX DESIGN GROUP

MONTGOMERY COUNTY, MARYLAND

R.A. SMITH NATIONAL

SOUTH FLORIDA WATER MANAGEMENT DISTRICT

February 8, 2012

Martin County Community Redevelopment Agency
Kevin Freeman, Director
2401 SE Monterey Road
Stuart, Florida 34996

Subject: **Stormwater Design Toolkit**

Dear Mr. Freeman:

The South Florida Water Management District (District) thanks you for the opportunity to review and comment on the Martin County Community Redevelopment Agency's (Agency) *Stormwater Design Toolkit – Sustainable Stormwater Update to the Community Redevelopment Area Stormwater Master Plan*.

The County completed the "Toolkit" document in support of the Stormwater Master Plan for the Community Redevelopment Areas (CRAs) of Martin County. This effort was undertaken to provide stormwater improvements while continuing to meet the goals of the Agency. The document is a collection of approaches and methodologies for treating and managing stormwater including an inventory of strategies which may be used independently and in combination to provide water quality treatment and stormwater attenuation.

The District supports the Agency's efforts to improve water quality with the production of this document. Our review found it to be a useful tool providing strategies that can assist in the stormwater management component of Agency projects.

The South Florida Water Management District commends the Martin County Community Redevelopment Agency for completing this document supporting the restoration and sustainability of water resources in South Florida.

Sincerely,

Hugo A. Carter, P.E.
Engineer Supervisor
Okeechobee Martin/St. Lucie Regulatory Division
South Florida Water Management District

HAC/hc

DISTRICT HEADQUARTERS: 3301 Gun Club Road, West Palm Beach, Florida 33406 • (561) 686-8800 • (800) 432-2045
Mailing Address: PO BOX 24680 West Palm Beach FL, 33416-4680
LOWER WEST COAST SERVICE CENTER: 2301 McGregor Boulevard, Fort Myers, FL 33901 • (239) 338-2929 • (800) 248-1201
OKEECHOBEE SERVICE CENTER: 205 North Parrott Avenue, Suite 201, Okeechobee, FL 34972 • (863) 462-5260 • (800) 250-4200
ORLANDO SERVICE CENTER: 1707 Orlando Central Parkway, Suite 200, Orlando FL 32809 • (407) 858-6100 • (800) 250-4250

www.ingramcontent.com/pod-product-compliance
Lightning Source LLC
Chambersburg PA
CBHW050731180526
45159CB00003B/1194